Markov Chains

Cambridge Series on Statistical and Probabilistic Mathematics

Editorial Board:
R. Gill (Utrecht)
B.D. Ripley (Oxford)
S. Ross (Berkeley)
M. Stein (Chicago)
D. Williams (Bath)

This series of high quality upper-division textbooks and expository monographs covers all areas of stochastic applicable mathematics. The topics range from pure and applied statistics to probability theory, operations research, mathematical programming, and optimzation. The books contain clear presentations of new developments in the field and also of the state of the art in classical methods. While emphasizing rigorous treatment of theoretical methods, the books contain important applications and discussions of new techniques made possible be advances in computational methods.

Markov Chains

J. R. Norris
University of Cambridge

CAMBRIDGE UNIVERSITY PRESS
Cambridge, New York, Melbourne, Madrid, Cape Town, Singapore, São Paulo, Delhi

Cambridge University Press
32 Avenue of the Americas, New York, NY 10013-2473, USA

www.cambridge.org
Information on this title: www.cambridge.org/9780521633963

© Cambridge University Press 1997

This publication is in copyright. Subject to statutory exception
and to the provisions of relevant collective licensing agreements,
no reproduction of any part may take place without the written
permission of Cambridge University Press.

First published 1997
First paperback edition 1998
15th printing 2009

Printed in the United States of America

A catalog record for this publication is available from the British Library.

Library of Congress Cataloging in Publication Data

Norris, J. R. (James R.)
Markov chains / J. R. Norris.
  p. cm. – (Cambridge series on statistical and probabilistic mathematics ; no. 2)
Includes bibliographical references and index.
ISBN 0-521-48181-3 (pbk.)
1. Markov processes.   I. Title.   II. Series.
QA274.7.N67   1997
519.2'33–dc20   96-31570
                                                                                    CIP

ISBN 978-0-521-48181-6 hardback
ISBN 978-0-521-63396-3 paperback

Cambridge University Press has no responsibility for the persistence or
accuracy of URLs for external or third-party Internet Web sites referred to in
this publication and does not guarantee that any content on such Web sites is,
or will remain, accurate or appropriate. Information regarding prices, travel
timetables, and other factual information given in this work are correct at
the time of first printing, but Cambridge University Press does not guarantee
the accuracy of such information thereafter.

For my parents

Contents

Preface	ix
Introduction	xiii
1. Discrete-time Markov chains	1
1.1 Definition and basic properties	1
1.2 Class structure	10
1.3 Hitting times and absorption probabilities	12
1.4 Strong Markov property	19
1.5 Recurrence and transience	24
1.6 Recurrence and transience of random walks	29
1.7 Invariant distributions	33
1.8 Convergence to equilibrium	40
1.9 Time reversal	47
1.10 Ergodic theorem	52
1.11 Appendix: recurrence relations	57
1.12 Appendix: asymptotics for $n!$	58
2. Continuous-time Markov chains I	60
2.1 Q-matrices and their exponentials	60
2.2 Continuous-time random processes	67
2.3 Some properties of the exponential distribution	70

2.4 Poisson processes	73
2.5 Birth processes	81
2.6 Jump chain and holding times	87
2.7 Explosion	90
2.8 Forward and backward equations	93
2.9 Non-minimal chains	103
2.10 Appendix: matrix exponentials	105
3. Continuous-time Markov chains II	**108**
3.1 Basic properties	108
3.2 Class structure	111
3.3 Hitting times and absorption probabilities	112
3.4 Recurrence and transience	114
3.5 Invariant distributions	117
3.6 Convergence to equilibrium	121
3.7 Time reversal	123
3.8 Ergodic theorem	125
4. Further theory	**128**
4.1 Martingales	128
4.2 Potential theory	134
4.3 Electrical networks	151
4.4 Brownian motion	159
5. Applications	**170**
5.1 Markov chains in biology	170
5.2 Queues and queueing networks	179
5.3 Markov chains in resource management	192
5.4 Markov decision processes	197
5.5 Markov chain Monte Carlo	206
6. Appendix: probability and measure	**217**
6.1 Countable sets and countable sums	217
6.2 Basic facts of measure theory	220
6.3 Probability spaces and expectation	222
6.4 Monotone convergence and Fubini's theorem	223
6.5 Stopping times and the strong Markov property	224
6.6 Uniqueness of probabilities and independence of σ-algebras	228
Further reading	**232**
Index	**234**

Preface

Markov chains are the simplest mathematical models for random phenomena evolving in time. Their simple structure makes it possible to say a great deal about their behaviour. At the same time, the class of Markov chains is rich enough to serve in many applications. This makes Markov chains the first and most important examples of random processes. Indeed, the whole of the mathematical study of random processes can be regarded as a generalization in one way or another of the theory of Markov chains.

This book is an account of the elementary theory of Markov chains, with applications. It was conceived as a text for advanced undergraduates or master's level students, and is developed from a course taught to undergraduates for several years. There are no strict prerequisites but it is envisaged that the reader will have taken a course in elementary probability. In particular, measure theory is not a prerequisite.

The first half of the book is based on lecture notes for the undergraduate course. Illustrative examples introduce many of the key ideas. Careful proofs are given throughout. There is a selection of exercises, which forms the basis of classwork done by the students, and which has been tested over several years. Chapter 1 deals with the theory of discrete-time Markov chains, and is the basis of all that follows. You must begin here. The material is quite straightforward and the ideas introduced permeate the whole book. The basic pattern of Chapter 1 is repeated in Chapter 3 for continuous-time chains, making it easy to follow the development by analogy. In between, Chapter 2 explains how to set up the theory of continuous-time chains, beginning with simple examples such as the Poisson process and chains with finite state space.

The second half of the book comprises three independent chapters intended to complement the first half. In some sections the style is a little more demanding. Chapter 4 introduces, in the context of elementary Markov chains, some of the ideas crucial to the advanced study of Markov processes, such as martingales, potentials, electrical networks and Brownian motion. Chapter 5 is devoted to applications, for example to population growth, mathematical genetics, queues and networks of queues, Markov decision processes and Monte Carlo simulation. Chapter 6 is an appendix to the main text, where we explain some of the basic notions of probability and measure used in the rest of the book and give careful proofs of the few points where measure theory is really needed.

The following paragraph is directed primarily at an instructor and assumes some familiarity with the subject. Overall, the book is more focused on the Markovian context than most other books dealing with the elementary theory of stochastic processes. I believe that this restriction in scope is desirable for the greater coherence and depth it allows. The treatment of discrete-time chains in Chapter 1 includes the calculation of transition probabilities, hitting probabilities, expected hitting times and invariant distributions. Also treated are recurrence and transience, convergence to equilibrium, reversibility, and the ergodic theorem for long-run averages. All the results are proved, exploiting to the full the probabilistic viewpoint. For example, we use excursions and the strong Markov property to obtain conditions for recurrence and transience, and convergence to equilibrium is proved by the coupling method. In Chapters 2 and 3 we proceed via the jump chain/holding time construction to treat all right-continuous, minimal continuous-time chains, and establish analogues of all the main results obtained for discrete time. No conditions of uniformly bounded rates are needed. The student has the option to take Chapter 3 first, to study the *properties* of continuous-time chains before the technically more demanding *construction*. We have left measure theory in the background, but the proofs are intended to be rigorous, or very easily made rigorous, when considered in measure-theoretic terms. Some further details are given in Chapter 6.

It is a pleasure to acknowledge the work of colleagues from which I have benefitted in preparing this book. The course on which it is based has evolved over many years and under many hands – I inherited parts of it from Martin Barlow and Chris Rogers. In recent years it has been given by Doug Kennedy and Colin Sparrow. Richard Gibbens, Geoffrey Grim-

Preface

mett, Frank Kelly and Gareth Roberts gave expert advice at various stages. Meena Lakshmanan, Violet Lo and David Rose pointed out many typos and ambiguities. Brian Ripley and David Williams made constructive suggestions for improvement of an early version.

I am especially grateful to David Tranah at Cambridge University Press for his suggestion to write the book and for his continuing support, and to Sarah Shea-Simonds who typeset the whole book with efficiency, precision and good humour.

Cambridge, 1996   James Norris

Introduction

This book is about a certain sort of random process. The characteristic property of this sort of process is that it retains *no memory* of where it has been in the past. This means that only the current state of the process can influence where it goes next. Such a process is called a *Markov process*. We shall be concerned exclusively with the case where the process can assume only a finite or countable set of states, when it is usual to refer it as a *Markov chain*.

Examples of Markov chains abound, as you will see throughout the book. What makes them important is that not only do Markov chains model many phenomena of interest, but also the lack of memory property makes it possible to predict how a Markov chain may behave, and to compute probabilities and expected values which quantify that behaviour. In this book we shall present general techniques for the analysis of Markov chains, together with many examples and applications. In this introduction we shall discuss a few very simple examples and preview some of the questions which the general theory will answer.

We shall consider chains both in *discrete time*
$$n \in \mathbb{Z}^+ = \{0, 1, 2, \dots\}$$
and *continuous time*
$$t \in \mathbb{R}^+ = [0, \infty).$$
The letters n, m, k will always denote integers, whereas t and s will refer to real numbers. Thus we write $(X_n)_{n \geq 0}$ for a discrete-time process and $(X_t)_{t \geq 0}$ for a continuous-time process.

xiv *Introduction*

Markov chains are often best described by diagrams, of which we now give some simple examples:

(i) (*Discrete time*)

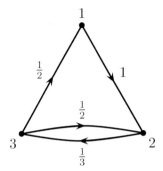

You move from state 1 to state 2 with probability 1. From state 3 you move either to 1 or to 2 with equal probability 1/2, and from 2 you jump to 3 with probability 1/3, otherwise stay at 2. We might have drawn a loop from 2 to itself with label 2/3. But since the total probability on jumping from 2 must equal 1, this does not convey any more information and we prefer to leave the loops out.

(ii) (*Continuous time*)

$$0 \xrightarrow{\lambda} 1$$

When in state 0 you wait for a random time with exponential distribution of parameter $\lambda \in (0, \infty)$, then jump to 1. Thus the density function of the waiting time T is given by

$$f_T(t) = \lambda e^{-\lambda t} \qquad \text{for } t \geq 0.$$

We write $T \sim E(\lambda)$ for short.

(iii) (*Continuous time*)

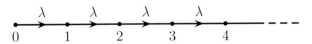

Here, when you get to 1 you do not stop but after another independent exponential time of parameter λ jump to 2, and so on. The resulting process is called the *Poisson process of rate* λ.

Introduction   xv

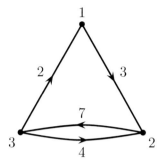

(iv) (*Continuous time*)

In state 3 you take two independent exponential times $T_1 \sim E(2)$ and $T_2 \sim E(4)$; if T_1 is the smaller you go to 1 after time T_1, and if T_2 is the smaller you go to 2 after time T_2. The rules for states 1 and 2 are as given in examples (ii) and (iii). It is a simple matter to show that the time spent in 3 is exponential of parameter $2 + 4 = 6$, and that the probability of jumping from 3 to 1 is $2/(2+4) = 1/3$. The details are given later.

(v) (*Discrete time*)

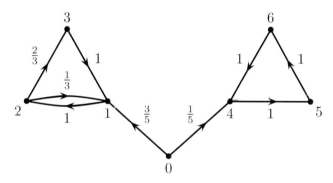

We use this example to anticipate some of the ideas discussed in detail in Chapter 1. The states may be partitioned into *communicating classes*, namely $\{0\}$, $\{1,2,3\}$ and $\{4,5,6\}$. Two of these classes are *closed*, meaning that you cannot escape. The closed classes here are *recurrent*, meaning that you return again and again to every state. The class $\{0\}$ is *transient*. The class $\{4,5,6\}$ is *periodic*, but $\{1,2,3\}$ is not. We shall show how to establish the following facts by solving some simple linear equations. You might like to try from first principles.

(a) Starting from 0, the probability of hitting 6 is 1/4.
(b) Starting from 1, the probability of hitting 3 is 1.
(c) Starting from 1, it takes on average three steps to hit 3.
(d) Starting from 1, the long-run proportion of time spent in 2 is 3/8.

Let us write $p_{ij}^{(n)}$ for the probability starting from i of being in state j after n steps. Then we have:

(e) $\lim_{n\to\infty} p_{01}^{(n)} = 9/32$;

(f) $p_{04}^{(n)}$ does not converge as $n \to \infty$;

(g) $\lim_{(n)\to\infty} p_{04}^{(3n)} = 1/124$.

1
Discrete-time Markov chains

This chapter is the foundation for all that follows. Discrete-time Markov chains are defined and their behaviour is investigated. For better orientation we now list the key theorems: these are Theorems 1.3.2 and 1.3.5 on hitting times, Theorem 1.4.2 on the strong Markov property, Theorem 1.5.3 characterizing recurrence and transience, Theorem 1.7.7 on invariant distributions and positive recurrence, Theorem 1.8.3 on convergence to equilibrium, Theorem 1.9.3 on reversibility, and Theorem 1.10.2 on long-run averages. Once you understand these you will understand the basic theory. Part of that understanding will come from familiarity with examples, so a large number are worked out in the text. Exercises at the end of each section are an important part of the exposition.

1.1 Definition and basic properties

Let I be a countable set. Each $i \in I$ is called a *state* and I is called the *state-space*. We say that $\lambda = (\lambda_i : i \in I)$ is a *measure* on I if $0 \leq \lambda_i < \infty$ for all $i \in I$. If in addition the *total mass* $\sum_{i \in I} \lambda_i$ equals 1, then we call λ a *distribution*. We work throughout with a probability space $(\Omega, \mathcal{F}, \mathbb{P})$. Recall that a *random variable* X with values in I is a function $X : \Omega \to I$. Suppose we set

$$\lambda_i = \mathbb{P}(X = i) = \mathbb{P}(\{\omega : X(\omega) = i\}).$$

Then λ defines a distribution, the *distribution of X*. We think of X as modelling a random state which takes the value i with probability λ_i. There is a brief review of some basic facts about countable sets and probability spaces in Chapter 6.

We say that a matrix $P = (p_{ij} : i, j \in I)$ is *stochastic* if every row $(p_{ij} : j \in I)$ is a distribution. There is a one-to-one correspondence between stochastic matrices P and the sort of diagrams described in the Introduction. Here are two examples:

$$P = \begin{pmatrix} 1-\alpha & \alpha \\ \beta & 1-\beta \end{pmatrix}$$

$$P = \begin{pmatrix} 0 & 1 & 0 \\ 0 & 1/2 & 1/2 \\ 1/2 & 0 & 1/2 \end{pmatrix}$$

We shall now formalize the rules for a Markov chain by a definition in terms of the corresponding matrix P. We say that $(X_n)_{n \geq 0}$ is a *Markov chain* with *initial distribution* λ and *transition matrix* P if

(i) X_0 has distribution λ;
(ii) for $n \geq 0$, conditional on $X_n = i$, X_{n+1} has distribution $(p_{ij} : j \in I)$ and is independent of $X_0, \ldots, X_{n-1}$.

More explicitly, these conditions state that, for $n \geq 0$ and $i_1, \ldots, i_{n+1} \in I$,

(i) $\mathbb{P}(X_0 = i_1) = \lambda_{i_1}$;
(ii) $\mathbb{P}(X_{n+1} = i_{n+1} \mid X_0 = i_1, \ldots, X_n = i_n) = p_{i_n i_{n+1}}$.

We say that $(X_n)_{n \geq 0}$ is $Markov(\lambda, P)$ for short. If $(X_n)_{0 \leq n \leq N}$ is a finite sequence of random variables satisfying (i) and (ii) for $n = 0, \ldots, N-1$, then we again say $(X_n)_{0 \leq n \leq N}$ is $Markov(\lambda, P)$.

It is in terms of properties (i) and (ii) that most real-world examples are seen to be Markov chains. But mathematically the following result appears to give a more comprehensive description, and it is the key to some later calculations.

Theorem 1.1.1. *A discrete-time random process $(X_n)_{0 \leq n \leq N}$ is $Markov(\lambda, P)$ if and only if for all $i_1, \ldots, i_N \in I$*

$$\mathbb{P}(X_0 = i_1, X_1 = i_2, \ldots, X_N = i_N) = \lambda_{i_1} p_{i_1 i_2} p_{i_2 i_3} \ldots p_{i_{N-1} i_N}. \quad (1.1)$$

1.1 Definition and basic properties

Proof. Suppose $(X_n)_{0 \leq n \leq N}$ is Markov(λ, P), then

$$\mathbb{P}(X_0 = i_1, X_1 = i_2, \ldots, X_N = i_N)$$
$$= \mathbb{P}(X_0 = i_1)\mathbb{P}(X_1 = i_2 \mid X_0 = i_1)$$
$$\ldots \mathbb{P}(X_N = i_N \mid X_0 = i_1, \ldots, X_{N-1} = i_{N-1})$$
$$= \lambda_{i_1} p_{i_1 i_2} \ldots p_{i_{N-1} i_N}.$$

On the other hand, if (1.1) holds for N, then by summing both sides over $i_N \in I$ and using $\sum_{j \in I} p_{ij} = 1$ we see that (1.1) holds for $N - 1$ and, by induction

$$\mathbb{P}(X_0 = i_1, X_1 = i_2, \ldots, X_n = i_n) = \lambda_{i_1} p_{i_1 i_2} \ldots p_{i_{n-1} i_n}$$

for all $n = 0, 1, \ldots, N$. In particular, $\mathbb{P}(X_0 = i_1) = \lambda_{i_1}$ and, for $n = 0, 1, \ldots, N-1$,

$$\mathbb{P}(X_{n+1} = i_{n+1} \mid X_0 = i_1, \ldots, X_n = i_n)$$
$$= \mathbb{P}(X_0 = i_1, \ldots, X_n = i_n, X_{n+1} = i_{n+1})/\mathbb{P}(X_0 = i_1, \ldots, X_n = i_n)$$
$$= p_{i_n i_{n+1}}.$$

So $(X_n)_{0 \leq n \leq N}$ is Markov(λ, P). □

The next result reinforces the idea that Markov chains have no memory. We write $\delta_i = (\delta_{ij} : j \in I)$ for the *unit mass* at i, where

$$\delta_{ij} = \begin{cases} 1 & \text{if } i = j \\ 0 & \text{otherwise.} \end{cases}$$

Theorem 1.1.2 (Markov property). *Let $(X_n)_{n \geq 0}$ be Markov(λ, P). Then, conditional on $X_m = i$, $(X_{m+n})_{n \geq 0}$ is Markov(δ_i, P) and is independent of the random variables $X_0, \ldots, X_m$.*

Proof. We have to show that for any event A determined by $X_0, \ldots, X_m$ we have

$$\mathbb{P}(\{X_m = i_m, \ldots, X_{m+n} = i_{m+n}\} \cap A \mid X_m = i)$$
$$= \delta_{i i_m} p_{i_m i_{m+1}} \ldots p_{i_{m+n-1} i_{m+n}} \mathbb{P}(A \mid X_m = i) \quad (1.2)$$

then the result follows by Theorem 1.1.1. First consider the case of elementary events

$$A = \{X_0 = i_1, \ldots, X_m = i_m\}.$$

In that case we have to show

$$\mathbb{P}(X_0 = i_1, \ldots, X_{m+n} = i_{m+n} \text{ and } i = i_m)/\mathbb{P}(X_m = i)$$
$$= \delta_{ii_m} p_{i_m i_{m+1}} \cdots p_{i_{m+n-1} i_{m+n}}$$
$$\times \mathbb{P}(X_0 = i_1, \ldots, X_m = i_m \text{ and } i = i_m)/\mathbb{P}(X_m = i)$$

which is true by Theorem 1.1.1. In general, any event A determined by $X_0, \ldots, X_m$ may be written as a countable disjoint union of elementary events

$$A = \bigcup_{k=1}^{\infty} A_k.$$

Then the desired identity (1.2) for A follows by summing up the corresponding identities for A_k. □

The remainder of this section addresses the following problem: *what is the probability that after n steps our Markov chain is in a given state?* First we shall see how the problem reduces to calculating entries in the nth power of the transition matrix. Then we shall look at some examples where this may be done explicitly.

We regard distributions and measures λ as row vectors whose components are indexed by I, just as P is a matrix whose entries are indexed by $I \times I$. When I is finite we will often label the states $1, 2, \ldots, N$; then λ will be an N-vector and P an $N \times N$-matrix. For these objects, matrix multiplication is a familiar operation. We extend matrix multiplication to the general case in the obvious way, defining a new measure λP and a new matrix P^2 by

$$(\lambda P)_j = \sum_{i \in I} \lambda_i p_{ij}, \quad (P^2)_{ik} = \sum_{j \in I} p_{ij} p_{jk}.$$

We define P^n similarly for any n. We agree that P^0 is the identity matrix I, where $(I)_{ij} = \delta_{ij}$. The context will make it clear when I refers to the state-space and when to the identity matrix. We write $p_{ij}^{(n)} = (P^n)_{ij}$ for the (i, j) entry in P^n.

In the case where $\lambda_i > 0$ we shall write $\mathbb{P}_i(A)$ for the conditional probability $\mathbb{P}(A \mid X_0 = i)$. By the Markov property at time $m = 0$, under $\mathbb{P}_i$, $(X_n)_{n \geq 0}$ is Markov(δ_i, P). So the behaviour of $(X_n)_{n \geq 0}$ under $\mathbb{P}_i$ does not depend on λ.

Theorem 1.1.3. *Let $(X_n)_{n \geq 0}$ be Markov(λ, P). Then, for all $n, m \geq 0$,*
  (i) $\mathbb{P}(X_n = j) = (\lambda P^n)_j$;
  (ii) $\mathbb{P}_i(X_n = j) = \mathbb{P}(X_{n+m} = j \mid X_m = i) = p_{ij}^{(n)}.$

1.1 Definition and basic properties

Proof. (i) By Theorem 1.1.1

$$\mathbb{P}(X_n = j) = \sum_{i_1 \in I} \cdots \sum_{i_{n-1} \in I} \mathbb{P}(X_0 = i_1, \ldots, X_{n-1} = i_{n-1}, X_n = j)$$

$$= \sum_{i_1 \in I} \cdots \sum_{i_{n-1} \in I} \lambda_{i_1} p_{i_1 i_2} \cdots p_{i_{n-1} j} = (\lambda P^n)_j.$$

(ii) By the Markov property, conditional on $X_m = i$, $(X_{m+n})_{n \geq 0}$ is Markov (δ_i, P), so we just take $\lambda = \delta_i$ in (i). □

In light of this theorem we call $p_{ij}^{(n)}$ the *n-step transition probability from i to j*. The following examples give some methods for calculating $p_{ij}^{(n)}$.

Example 1.1.4

The most general two-state chain has transition matrix of the form

$$P = \begin{pmatrix} 1 - \alpha & \alpha \\ \beta & 1 - \beta \end{pmatrix}$$

and is represented by the following diagram:

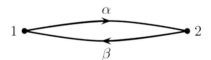

We exploit the relation $P^{n+1} = P^n P$ to write

$$p_{11}^{(n+1)} = p_{12}^{(n)} \beta + p_{11}^{(n)} (1 - \alpha).$$

We also know that $p_{11}^{(n)} + p_{12}^{(n)} = \mathbb{P}_1(X_n = 1 \text{ or } 2) = 1$, so by eliminating $p_{12}^{(n)}$ we get a recurrence relation for $p_{11}^{(n)}$:

$$p_{11}^{(n+1)} = (1 - \alpha - \beta) p_{11}^{(n)} + \beta, \quad p_{11}^{(0)} = 1.$$

This has a unique solution (see Section 1.11):

$$p_{11}^{(n)} = \begin{cases} \dfrac{\beta}{\alpha + \beta} + \dfrac{\alpha}{\alpha + \beta} (1 - \alpha - \beta)^n & \text{for } \alpha + \beta > 0 \\ 1 & \text{for } \alpha + \beta = 0. \end{cases}$$

Example 1.1.5 (Virus mutation)

Suppose a virus can exist in N different strains and in each generation either stays the same, or with probability α mutates to another strain, which is chosen at random. What is the probability that the strain in the nth generation is the same as that in the 0th?

We could model this process as an N-state chain, with $N \times N$ transition matrix P given by

$$p_{ii} = 1 - \alpha, \quad p_{ij} = \alpha/(N-1) \quad \text{for } i \neq j.$$

Then the answer we want would be found by computing $p_{11}^{(n)}$. In fact, in this example there is a much simpler approach, which relies on exploiting the symmetry present in the mutation rules.

At any time a transition is made from the initial state to another with probability α, and a transition from another state to the initial state with probability $\alpha/(N-1)$. Thus we have a two-state chain with diagram

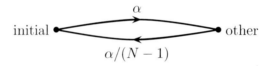

and by putting $\beta = \alpha/(N-1)$ in Example 1.1.4 we find that the desired probability is

$$\frac{1}{N} + \left(1 - \frac{1}{N}\right)\left(1 - \frac{\alpha N}{N-1}\right)^n.$$

Beware that in examples having less symmetry, this sort of lumping together of states may not produce a Markov chain.

Example 1.1.6

Consider the three-state chain with diagram

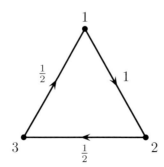

and transition matrix

$$P = \begin{pmatrix} 0 & 1 & 0 \\ 0 & \frac{1}{2} & \frac{1}{2} \\ \frac{1}{2} & 0 & \frac{1}{2} \end{pmatrix}.$$

The problem is to find a general formula for $p_{11}^{(n)}$.

First we compute the eigenvalues of P by writing down its characteristic equation

$$0 = \det(x - P) = x(x - \tfrac{1}{2})^2 - \tfrac{1}{4} = \tfrac{1}{4}(x-1)(4x^2 + 1).$$

The eigenvalues are $1, i/2, -i/2$ and from this we deduce that $p_{11}^{(n)}$ has the form

$$p_{11}^{(n)} = a + b\left(\frac{i}{2}\right)^n + c\left(-\frac{i}{2}\right)^n$$

for some constants a, b and c. (The justification comes from linear algebra: having distinct eigenvalues, P is diagonalizable, that is, for some invertible matrix U we have

$$P = U \begin{pmatrix} 1 & 0 & 0 \\ 0 & i/2 & 0 \\ 0 & 0 & -i/2 \end{pmatrix} U^{-1}$$

and hence

$$P^n = U \begin{pmatrix} 1 & 0 & 0 \\ 0 & (i/2)^n & 0 \\ 0 & 0 & (-i/2)^n \end{pmatrix} U^{-1}$$

which forces $p_{11}^{(n)}$ to have the form claimed.) The answer we want is real and

$$\left(\pm \frac{i}{2}\right)^n = \left(\frac{1}{2}\right)^n e^{\pm i n\pi/2} = \left(\frac{1}{2}\right)^n \left(\cos\frac{n\pi}{2} \pm i \sin\frac{n\pi}{2}\right)$$

so it makes sense to rewrite $p_{11}^{(n)}$ in the form

$$p_{11}^{(n)} = \alpha + \left(\frac{1}{2}\right)^n \left\{\beta \cos\frac{n\pi}{2} + \gamma \sin\frac{n\pi}{2}\right\}$$

for constants α, β and γ. The first few values of $p_{11}^{(n)}$ are easy to write down, so we get equations to solve for α, β and γ:

$$1 = p_{11}^{(0)} = \alpha + \beta$$
$$0 = p_{11}^{(1)} = \alpha + \tfrac{1}{2}\gamma$$
$$0 = p_{11}^{(2)} = \alpha - \tfrac{1}{4}\beta$$

so $\alpha = 1/5$, $\beta = 4/5$, $\gamma = -2/5$ and
$$p_{11}^{(n)} = \frac{1}{5} + \left(\frac{1}{2}\right)^n \left\{\frac{4}{5}\cos\frac{n\pi}{2} - \frac{2}{5}\sin\frac{n\pi}{2}\right\}.$$

More generally, the following method may in principle be used to find a formula for $p_{ij}^{(n)}$ for any M-state chain and any states i and j.

(i) Compute the eigenvalues $\lambda_1, \ldots, \lambda_M$ of P by solving the characteristic equation.

(ii) If the eigenvalues are distinct then $p_{ij}^{(n)}$ has the form
$$p_{ij}^{(n)} = a_1 \lambda_1^n + \ldots + a_M \lambda_M^n$$
for some constants $a_1, \ldots, a_M$ (depending on i and j). If an eigenvalue λ is repeated (once, say) then the general form includes the term $(an + b)\lambda^n$.

(iii) As roots of a polynomial with real coefficients, complex eigenvalues will come in conjugate pairs and these are best written using sine and cosine, as in the example.

Exercises

1.1.1 Let $B_1, B_2, \ldots$ be disjoint events with $\bigcup_{n=1}^{\infty} B_n = \Omega$. Show that if A is another event and $\mathbb{P}(A|B_n) = p$ for all n then $\mathbb{P}(A) = p$.

Deduce that if X and Y are discrete random variables then the following are equivalent:

(a) X and Y are independent;

(b) the conditional distribution of X given $Y = y$ is independent of y.

1.1.2 Suppose that $(X_n)_{n \geq 0}$ is Markov (λ, P). If $Y_n = X_{kn}$, show that $(Y_n)_{n \geq 0}$ is Markov (λ, P^k).

1.1.3 Let X_0 be a random variable with values in a countable set I. Let $Y_1, Y_2, \ldots$ be a sequence of independent random variables, uniformly distributed on $[0, 1]$. Suppose we are given a function
$$G : I \times [0, 1] \to I$$
and define inductively
$$X_{n+1} = G(X_n, Y_{n+1}).$$

Show that $(X_n)_{n \geq 0}$ is a Markov chain and express its transition matrix P in terms of G. Can all Markov chains be realized in this way? How would you simulate a Markov chain using a computer?

Suppose now that $Z_0, Z_1, \ldots$ are independent, identically distributed random variables such that $Z_i = 1$ with probability p and $Z_i = 0$ with probability $1 - p$. Set $S_0 = 0$, $S_n = Z_1 + \ldots + Z_n$. In each of the following cases determine whether $(X_n)_{n \geq 0}$ is a Markov chain:

(a) $X_n = Z_n$, (b) $X_n = S_n$,
(c) $X_n = S_0 + \ldots + S_n$, (d) $X_n = (S_n, S_0 + \ldots + S_n)$.

In the cases where $(X_n)_{n \geq 0}$ is a Markov chain find its state-space and transition matrix, and in the cases where it is not a Markov chain give an example where $P(X_{n+1} = i | X_n = j, X_{n-1} = k)$ is not independent of k.

1.1.4 A flea hops about at random on the vertices of a triangle, with all jumps equally likely. Find the probability that after n hops the flea is back where it started.

A second flea also hops about on the vertices of a triangle, but this flea is twice as likely to jump clockwise as anticlockwise. What is the probability that after n hops this second flea is back where it started? [Recall that $e^{\pm i\pi/6} = \sqrt{3}/2 \pm i/2$.]

1.1.5 A die is 'fixed' so that each time it is rolled the score cannot be the same as the preceding score, all other scores having probability $1/5$. If the first score is 6, what is the probability p that the nth score is 6? What is the probability that the nth score is 1?

Suppose now that a new die is produced which cannot score one greater (mod 6) than the preceding score, all other scores having equal probability. By considering the relationship between the two dice find the value of p for the new die.

1.1.6 An octopus is trained to choose object A from a pair of objects A, B by being given repeated trials in which it is shown both and is rewarded with food if it chooses A. The octopus may be in one of three states of mind: in state 1 it cannot remember which object is rewarded and is equally likely to choose either; in state 2 it remembers and chooses A but may forget again; in state 3 it remembers and chooses A and never forgets. After each trial it may change its state of mind according to the transition matrix

$$\begin{array}{cccc} \text{State 1} & \frac{1}{2} & \frac{1}{2} & 0 \\ \text{State 2} & \frac{1}{2} & \frac{1}{12} & \frac{5}{12} \\ \text{State 3} & 0 & 0 & 1. \end{array}$$

It is in state 1 before the first trial. What is the probablity that it is in state 1 just before the $(n+1)$th trial ? What is the probability $P_{n+1}(A)$ that it chooses A on the $(n + 1)$th trial ?

Someone suggests that the record of successive choices (a sequence of As and Bs) might arise from a two-state Markov chain with constant transition probabilities. Discuss, with reference to the value of $P_{n+1}(A)$ that you have found, whether this is possible.

1.1.7 Let $(X_n)_{n \geq 0}$ be a Markov chain on $\{1, 2, 3\}$ with transition matrix

$$P = \begin{pmatrix} 0 & 1 & 0 \\ 0 & 2/3 & 1/3 \\ p & 1-p & 0 \end{pmatrix}.$$

Calculate $\mathbb{P}(X_n = 1 | X_0 = 1)$ in each of the following cases: (a) $p = 1/16$, (b) $p = 1/6$, (c) $p = 1/12$.

1.2 Class structure

It is sometimes possible to break a Markov chain into smaller pieces, each of which is relatively easy to understand, and which together give an understanding of the whole. This is done by identifying the communicating classes of the chain.

We say that i *leads to* j and write $i \to j$ if

$$\mathbb{P}_i(X_n = j \text{ for some } n \geq 0) > 0.$$

We say i *communicates with* j and write $i \leftrightarrow j$ if both $i \to j$ and $j \to i$.

Theorem 1.2.1. *For distinct states i and j the following are equivalent:*
  (i) $i \to j$;
  (ii) $p_{i_1 i_2} p_{i_2 i_3} \ldots p_{i_{n-1} i_n} > 0$ *for some states $i_1, i_2, \ldots, i_n$ with $i_1 = i$ and $i_n = j$;*
  (iii) $p_{ij}^{(n)} > 0$ *for some $n \geq 0$.*

Proof. Observe that

$$p_{ij}^{(n)} \leq \mathbb{P}_i(X_n = j \text{ for some } n \geq 0) \leq \sum_{n=0}^{\infty} p_{ij}^{(n)}$$

which proves the equivalence of (i) and (iii). Also

$$p_{ij}^{(n)} = \sum_{i_2, \ldots, i_{n-1}} p_{i i_2} p_{i_2 i_3} \ldots p_{i_{n-1} j}$$

so that (ii) and (iii) are equivalent. □

1.3 Hitting times and absorption probabilities

It is clear from (ii) that $i \to j$ and $j \to k$ imply $i \to k$. Also $i \to i$ for any state i. So $\leftrightarrow$ satisfies the conditions for an equivalence relation on I, and thus partitions I into *communicating classes*. We say that a class C is *closed* if

$$i \in C, i \to j \quad \text{imply } j \in C.$$

Thus a closed class is one from which there is no escape. A state i is *absorbing* if $\{i\}$ is a closed class. The smaller pieces referred to above are these communicating classes. A chain or transition matrix P where I is a single class is called *irreducible*.

As the following example makes clear, when one can draw the diagram, the class structure of a chain is very easy to find.

Example 1.2.2

Find the communicating classes associated to the stochastic matrix

$$P = \begin{pmatrix} \frac{1}{2} & \frac{1}{2} & 0 & 0 & 0 & 0 \\ 0 & 0 & 1 & 0 & 0 & 0 \\ \frac{1}{3} & 0 & 0 & \frac{1}{3} & \frac{1}{3} & 0 \\ 0 & 0 & 0 & \frac{1}{2} & \frac{1}{2} & 0 \\ 0 & 0 & 0 & 0 & 0 & 1 \\ 0 & 0 & 0 & 0 & 1 & 0 \end{pmatrix}.$$

The solution is obvious from the diagram

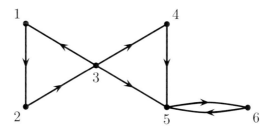

the classes being $\{1,2,3\}$, $\{4\}$ and $\{5,6\}$, with only $\{5,6\}$ being closed.

Exercises

1.2.1 Identify the communicating classes of the following transition matrix:

$$P = \begin{pmatrix} \frac{1}{2} & 0 & 0 & 0 & \frac{1}{2} \\ 0 & \frac{1}{2} & 0 & \frac{1}{2} & 0 \\ 0 & 0 & 1 & 0 & 0 \\ 0 & \frac{1}{4} & \frac{1}{4} & \frac{1}{4} & \frac{1}{4} \\ \frac{1}{2} & 0 & 0 & 0 & \frac{1}{2} \end{pmatrix}.$$

Which classes are closed?

1.2.2 Show that every transition matrix on a finite state-space has at least one closed communicating class. Find an example of a transition matrix with no closed communicating class.

1.3 Hitting times and absorption probabilities

Let $(X_n)_{n\geq 0}$ be a Markov chain with transition matrix P. The *hitting time* of a subset A of I is the random variable $H^A : \Omega \to \{0, 1, 2, \dots\} \cup \{\infty\}$ given by
$$H^A(\omega) = \inf\{n \geq 0 : X_n(\omega) \in A\}$$
where we agree that the infimum of the empty set $\emptyset$ is ∞. The probability starting from i that $(X_n)_{n\geq 0}$ ever hits A is then
$$h_i^A = \mathbb{P}_i(H^A < \infty).$$
When A is a closed class, h_i^A is called the *absorption probability*. The mean time taken for $(X_n)_{n\geq 0}$ to reach A is given by
$$k_i^A = \mathbb{E}_i(H^A) = \sum_{n<\infty} n\mathbb{P}(H^A = n) + \infty\mathbb{P}(H^A = \infty).$$
We shall often write less formally
$$h_i^A = \mathbb{P}_i(\text{hit } A), \quad k_i^A = \mathbb{E}_i(\text{time to hit } A).$$
Remarkably, these quantities can be calculated explicitly by means of certain linear equations associated with the transition matrix P. Before we give the general theory, here is a simple example.

Example 1.3.1

Consider the chain with the following diagram:

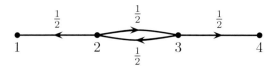

Starting from 2, what is the probability of absorption in 4? How long does it take until the chain is absorbed in 1 or 4?

Introduce
$$h_i = \mathbb{P}_i(\text{hit } 4), \quad k_i = \mathbb{E}_i(\text{time to hit } \{1,4\}).$$

1.3 Hitting times and absorption probabilities

Clearly, $h_1 = 0$, $h_4 = 1$ and $k_1 = k_4 = 0$. Suppose now that we start at 2, and consider the situation after making one step. With probability 1/2 we jump to 1 and with probability 1/2 we jump to 3. So

$$h_2 = \tfrac{1}{2}h_1 + \tfrac{1}{2}h_3, \quad k_2 = 1 + \tfrac{1}{2}k_1 + \tfrac{1}{2}k_3.$$

The 1 appears in the second formula because we count the time for the first step. Similarly,

$$h_3 = \tfrac{1}{2}h_2 + \tfrac{1}{2}h_4, \quad k_3 = 1 + \tfrac{1}{2}k_2 + \tfrac{1}{2}k_4.$$

Hence

$$h_2 = \tfrac{1}{2}h_3 = \tfrac{1}{2}(\tfrac{1}{2}h_2 + \tfrac{1}{2}),$$
$$k_2 = 1 + \tfrac{1}{2}k_3 = 1 + \tfrac{1}{2}(1 + \tfrac{1}{2}k_2).$$

So, starting from 2, the probability of hitting 4 is 1/3 and the mean time to absorption is 2. Note that in writing down the first equations for h_2 and k_2 we made implicit use of the Markov property, in assuming that the chain begins afresh from its new position after the first jump. Here is a general result for hitting probabilities.

Theorem 1.3.2. *The vector of hitting probabilities $h^A = (h_i^A : i \in I)$ is the minimal non-negative solution to the system of linear equations*

$$\begin{cases} h_i^A = 1 & \text{for } i \in A \\ h_i^A = \sum_{j \in I} p_{ij} h_j^A & \text{for } i \notin A. \end{cases} \quad (1.3)$$

(Minimality means that if $x = (x_i : i \in I)$ is another solution with $x_i \geq 0$ for all i, then $x_i \geq h_i$ for all i.)

Proof. First we show that h^A satisfies (1.3). If $X_0 = i \in A$, then $H^A = 0$, so $h_i^A = 1$. If $X_0 = i \notin A$, then $H^A \geq 1$, so by the Markov property

$$\mathbb{P}_i(H^A < \infty \mid X_1 = j) = \mathbb{P}_j(H^A < \infty) = h_j^A$$

and

$$h_i^A = \mathbb{P}_i(H^A < \infty) = \sum_{j \in I} \mathbb{P}_i(H^A < \infty, X_1 = j)$$
$$= \sum_{j \in I} \mathbb{P}_i(H^A < \infty \mid X_1 = j)\mathbb{P}_i(X_1 = j) = \sum_{j \in I} p_{ij} h_j^A.$$

Suppose now that $x = (x_i : i \in I)$ is any solution to (1.3). Then $h_i^A = x_i = 1$ for $i \in A$. Suppose $i \notin A$, then

$$x_i = \sum_{j \in I} p_{ij} x_j = \sum_{j \in A} p_{ij} + \sum_{j \notin A} p_{ij} x_j.$$

Substitute for x_j to obtain

$$x_i = \sum_{j \in A} p_{ij} + \sum_{j \notin A} p_{ij} \left(\sum_{k \in A} p_{jk} + \sum_{k \notin A} p_{jk} x_k \right)$$
$$= \mathbb{P}_i(X_1 \in A) + \mathbb{P}_i(X_1 \notin A, X_2 \in A) + \sum_{j \notin A} \sum_{k \notin A} p_{ij} p_{jk} x_k.$$

By repeated substitution for x in the final term we obtain after n steps

$$x_i = \mathbb{P}_i(X_1 \in A) + \ldots + \mathbb{P}_i(X_1 \notin A, \ldots, X_{n-1} \notin A, X_n \in A)$$
$$+ \sum_{j_1 \notin A} \cdots \sum_{j_n \notin A} p_{ij_1} p_{j_1 j_2} \cdots p_{j_{n-1} j_n} x_{j_n}.$$

Now if x is non-negative, so is the last term on the right, and the remaining terms sum to $\mathbb{P}_i(H^A \leq n)$. So $x_i \geq \mathbb{P}_i(H^A \leq n)$ for all n and then

$$x_i \geq \lim_{n \to \infty} \mathbb{P}_i(H^A \leq n) = \mathbb{P}_i(H^A < \infty) = h_i. \qquad \square$$

Example 1.3.1 (continued)

The system of linear equations (1.3) for $h = h^{\{4\}}$ are given here by

$$h_4 = 1,$$
$$h_2 = \tfrac{1}{2} h_1 + \tfrac{1}{2} h_3, \ h_3 = \tfrac{1}{2} h_2 + \tfrac{1}{2} h_4$$

so that

$$h_2 = \tfrac{1}{2} h_1 + \tfrac{1}{2}(\tfrac{1}{2} h_2 + \tfrac{1}{2})$$

and

$$h_2 = \tfrac{1}{3} + \tfrac{2}{3} h_1, \ h_3 = \tfrac{2}{3} + \tfrac{1}{3} h_1.$$

The value of h_1 is not determined by the system (1.3), but the minimality condition now makes us take $h_1 = 0$, so we recover $h_2 = 1/3$ as before. Of course, the extra boundary condition $h_1 = 0$ was obvious from the beginning

1.3 Hitting times and absorption probabilities

so we built it into our system of equations and did not have to worry about minimal non-negative solutions.

In cases where the state-space is infinite it may not be possible to write down a corresponding extra boundary condition. Then, as we shall see in the next examples, the minimality condition is essential.

Example 1.3.3 (Gamblers' ruin)

Consider the Markov chain with diagram

where $0 < p = 1 - q < 1$. The transition probabilities are

$$p_{00} = 1,$$
$$p_{i,i-1} = q, \; p_{i,i+1} = p \quad \text{for } i = 1, 2, \ldots.$$

Imagine that you enter a casino with a fortune of £i and gamble, £1 at a time, with probability p of doubling your stake and probability q of losing it. The resources of the casino are regarded as infinite, so there is no upper limit to your fortune. But what is the probability that you leave broke?

Set $h_i = \mathbb{P}_i(\text{hit } 0)$, then h is the minimal non-negative solution to

$$h_0 = 1,$$
$$h_i = p h_{i+1} + q h_{i-1}, \quad \text{for } i = 1, 2, \ldots.$$

If $p \neq q$ this recurrence relation has a general solution

$$h_i = A + B \left(\frac{q}{p}\right)^i.$$

(See Section 1.11.) If $p < q$, which is the case in most successful casinos, then the restriction $0 \leq h_i \leq 1$ forces $B = 0$, so $h_i = 1$ for all i. If $p > q$, then since $h_0 = 1$ we get a family of solutions

$$h_i = \left(\frac{q}{p}\right)^i + A\left(1 - \left(\frac{q}{p}\right)^i\right);$$

for a non-negative solution we must have $A \geq 0$, so the minimal non-negative solution is $h_i = (q/p)^i$. Finally, if $p = q$ the recurrence relation has a general solution

$$h_i = A + Bi$$

and again the restriction $0 \leq h_i \leq 1$ forces $B = 0$, so $h_i = 1$ for all i. Thus, even if you find a fair casino, you are certain to end up broke. This apparent paradox is called gamblers' ruin.

Example 1.3.4 (Birth-and-death chain)

Consider the Markov chain with diagram

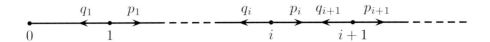

where, for $i = 1, 2, \ldots$, we have $0 < p_i = 1 - q_i < 1$. As in the preceding example, 0 is an absorbing state and we wish to calculate the absorption probability starting from i. But here we allow p_i and q_i to depend on i.

Such a chain may serve as a model for the size of a population, recorded each time it changes, p_i being the probability that we get a birth before a death in a population of size i. Then $h_i = \mathbb{P}_i(\text{hit } 0)$ is the extinction probability starting from i.

We write down the usual system of equations

$$h_0 = 1,$$
$$h_i = p_i h_{i+1} + q_i h_{i-1}, \quad \text{for } i = 1, 2, \ldots.$$

This recurrence relation has variable coefficients so the usual technique fails. But consider $u_i = h_{i-1} - h_i$, then $p_i u_{i+1} = q_i u_i$, so

$$u_{i+1} = \left(\frac{q_i}{p_i}\right) u_i = \left(\frac{q_i q_{i-1} \cdots q_1}{p_i p_{i-1} \cdots p_1}\right) u_1 = \gamma_i u_1$$

where the final equality defines γ_i. Then

$$u_1 + \ldots + u_i = h_0 - h_i$$

so

$$h_i = 1 - A(\gamma_0 + \ldots + \gamma_{i-1})$$

where $A = u_1$ and $\gamma_0 = 1$. At this point A remains to be determined. In the case $\sum_{i=0}^{\infty} \gamma_i = \infty$, the restriction $0 \leq h_i \leq 1$ forces $A = 0$ and $h_i = 1$ for all i. But if $\sum_{i=0}^{\infty} \gamma_i < \infty$ then we can take $A > 0$ so long as

$$1 - A(\gamma_0 + \ldots + \gamma_{i-1}) \geq 0 \quad \text{for all } i.$$

Thus the minimal non-negative solution occurs when $A = \left(\sum_{i=0}^{\infty} \gamma_i\right)^{-1}$ and then
$$h_i = \sum_{j=i}^{\infty} \gamma_j \bigg/ \sum_{j=0}^{\infty} \gamma_j.$$

In this case, for $i = 1, 2, \ldots$, we have $h_i < 1$, so the population survives with positive probability.

Here is the general result on mean hitting times. Recall that $k_i^A = \mathbb{E}_i(H^A)$, where H^A is the first time $(X_n)_{n \geq 0}$ hits A. We use the notation 1_B for the indicator function of B, so, for example, $1_{X_1=j}$ is the random variable equal to 1 if $X_1 = j$ and equal to 0 otherwise.

Theorem 1.3.5. *The vector of mean hitting times $k^A = (k^A : i \in I)$ is the minimal non-negative solution to the system of linear equations*
$$\begin{cases} k_i^A = 0 & \text{for } i \in A \\ k_i^A = 1 + \sum_{j \notin A} p_{ij} k_j^A & \text{for } i \notin A. \end{cases} \quad (1.4)$$

Proof. First we show that k^A satisfies (1.4). If $X_0 = i \in A$, then $H^A = 0$, so $k_i^A = 0$. If $X_0 = i \notin A$, then $H^A \geq 1$, so, by the Markov property,
$$\mathbb{E}_i(H^A \mid X_1 = j) = 1 + \mathbb{E}_j(H^A)$$
and
$$k_i^A = \mathbb{E}_i(H^A) = \sum_{j \in I} \mathbb{E}_i(H^A 1_{X_1=j})$$
$$= \sum_{j \in I} \mathbb{E}_i(H^A \mid X_1 = j) \mathbb{P}_i(X_1 = j) = 1 + \sum_{j \notin A} p_{ij} k_j^A.$$

Suppose now that $y = (y_i : i \in I)$ is any solution to (1.4). Then $k_i^A = y_i = 0$ for $i \in A$. If $i \notin A$, then
$$y_i = 1 + \sum_{j \notin A} p_{ij} y_j$$
$$= 1 + \sum_{j \notin A} p_{ij} \left(1 + \sum_{k \notin A} p_{jk} y_k \right)$$
$$= \mathbb{P}_i(H^A \geq 1) + \mathbb{P}_i(H^A \geq 2) + \sum_{j \notin A} \sum_{k \notin A} p_{ij} p_{jk} y_k.$$

By repeated substitution for y in the final term we obtain after n steps
$$y_i = \mathbb{P}_i(H^A \geq 1) + \ldots + \mathbb{P}_i(H^A \geq n) + \sum_{j_1 \notin A} \cdots \sum_{j_n \notin A} p_{ij_1} p_{j_1 j_2} \cdots p_{j_{n-1} j_n} y_{j_n}.$$

So, if y is non-negative,
$$y_i \geq \mathbb{P}_i(H^A \geq 1) + \ldots + \mathbb{P}_i(H^A \geq n)$$
and, letting $n \to \infty$,
$$y_i \geq \sum_{n=1}^{\infty} \mathbb{P}_i(H^A \geq n) = \mathbb{E}_i(H^A) = k_i^A. \qquad \square$$

Exercises

1.3.1 Prove the claims (a), (b) and (c) made in example (v) of the Introduction.

1.3.2 A gambler has £2 and needs to increase it to £10 in a hurry. He can play a game with the following rules: a fair coin is tossed; if a player bets on the right side, he wins a sum equal to his stake, and his stake is returned; otherwise he loses his stake. The gambler decides to use a bold strategy in which he stakes all his money if he has £5 or less, and otherwise stakes just enough to increase his capital, if he wins, to £10.

Let $X_0 = 2$ and let X_n be his capital after n throws. Prove that the gambler will achieve his aim with probability $1/5$.

What is the expected number of tosses until the gambler either achieves his aim or loses his capital?

1.3.3 A simple game of 'snakes and ladders' is played on a board of nine squares.

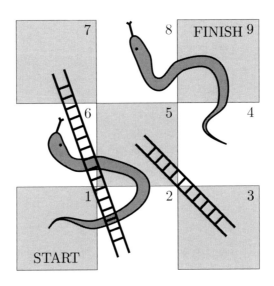

At each turn a player tosses a fair coin and advances one or two places according to whether the coin lands heads or tails. If you land at the foot of a ladder you climb to the top, but if you land at the head of a snake you slide down to the tail. How many turns on average does it take to complete the game?

What is the probability that a player who has reached the middle square will complete the game without slipping back to square 1?

1.3.4 Let $(X_n)_{n \geq 0}$ be a Markov chain on $\{0, 1, \dots\}$ with transition probabilities given by

$$p_{01} = 1, \quad p_{i,i+1} + p_{i,i-1} = 1, \quad p_{i,i+1} = \left(\frac{i+1}{i}\right)^2 p_{i,i-1}, \quad i \geq 1.$$

Show that if $X_0 = 0$ then the probability that $X_n \geq 1$ for all $n \geq 1$ is $6/\pi^2$.

1.4 Strong Markov property

In Section 1.1 we proved the Markov property. This says that for each time m, conditional on $X_m = i$, the process after time m begins afresh from i. Suppose, instead of conditioning on $X_m = i$, we simply waited for the process to hit state i, at some random time H. What can one say about the process after time H? What if we replaced H by a more general random time, for example $H - 1$? In this section we shall identify a class of random times at which a version of the Markov property does hold. This class will include H but not $H - 1$; after all, the process after time $H - 1$ jumps straight to i, so it does not simply begin afresh.

A random variable $T : \Omega \to \{0, 1, 2, \dots\} \cup \{\infty\}$ is called a *stopping time* if the event $\{T = n\}$ depends only on $X_0, X_1, \dots, X_n$ for $n = 0, 1, 2, \dots$. Intuitively, by watching the process, you know at the time when T occurs. If asked to stop at T, you know when to stop.

Examples 1.4.1

(a) The *first passage time*

$$T_j = \inf\{n \geq 1 : X_n = j\}$$

is a stopping time because

$$\{T_j = n\} = \{X_1 \neq j, \dots, X_{n-1} \neq j, X_n = j\}.$$

(b) The first hitting time H^A of Section 1.3 is a stopping time because

$$\{H^A = n\} = \{X_0 \notin A, \dots, X_{n-1} \notin A, X_n \in A\}.$$

(c) The *last exit time*

$$L^A = \sup\{n \geq 0 : X_n \in A\}$$

is not in general a stopping time because the event $\{L^A = n\}$ depends on whether $(X_{n+m})_{m \geq 1}$ visits A or not.

We shall show that the Markov property holds at stopping times. The crucial point is that, if T is a stopping time and $B \subseteq \Omega$ is determined by $X_0, X_1, \ldots, X_T$, then $B \cap \{T = m\}$ is determined by $X_0, X_1, \ldots, X_m$, for all $m = 0, 1, 2, \ldots$.

Theorem 1.4.2 (Strong Markov property). Let $(X_n)_{n \geq 0}$ be Markov(λ, P) and let T be a stopping time of $(X_n)_{n \geq 0}$. Then, conditional on $T < \infty$ and $X_T = i$, $(X_{T+n})_{n \geq 0}$ is Markov(δ_i, P) and independent of $X_0, X_1, \ldots, X_T$.

Proof. If B is an event determined by $X_0, X_1, \ldots, X_T$, then $B \cap \{T = m\}$ is determined by $X_0, X_1, \ldots, X_m$, so, by the Markov property at time m

$$\mathbb{P}(\{X_T = j_0, X_{T+1} = j_1, \ldots, X_{T+n} = j_n\} \cap B \cap \{T = m\} \cap \{X_T = i\})$$
$$= \mathbb{P}_i(X_0 = j_0, X_1 = j_1, \ldots, X_n = j_n)\mathbb{P}(B \cap \{T = m\} \cap \{X_T = i\})$$

where we have used the condition $T = m$ to replace m by T. Now sum over $m = 0, 1, 2, \ldots$ and divide by $\mathbb{P}(T < \infty, X_T = i)$ to obtain

$$\mathbb{P}(\{X_T = j_0, X_{T+1} = j_1, \ldots, X_{T+n} = j_n\} \cap B \mid T < \infty, X_T = i)$$
$$= \mathbb{P}_i(X_0 = j_0, X_1 = j_1, \ldots, X_n = j_n)\mathbb{P}(B \mid T < \infty, X_T = i). \qquad \square$$

The following example uses the strong Markov property to get more information on the hitting times of the chain considered in Example 1.3.3.

Example 1.4.3

Consider the Markov chain $(X_n)_{n \geq 0}$ with diagram

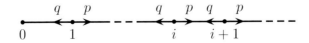

where $0 < p = 1 - q < 1$. We know from Example 1.3.3 the probability of hitting 0 starting from 1. Here we obtain the complete distribution of the time to hit 0 starting from 1 in terms of its probability generating function. Set
$$H_j = \inf\{n \geq 0 : X_n = j\}$$
and, for $0 \leq s < 1$
$$\phi(s) = \mathbb{E}_1(s^{H_0}) = \sum_{n < \infty} s^n \mathbb{P}_1(H_0 = n).$$

Suppose we start at 2. Apply the strong Markov property at H_1 to see that under $\mathbb{P}_2$, conditional on $H_1 < \infty$, we have $H_0 = H_1 + \widetilde{H}_0$, where $\widetilde{H}_0$, the time taken after H_1 to get to 0, is independent of H_1 and has the (unconditioned) distribution of H_1. So
$$\mathbb{E}_2(s^{H_0}) = \mathbb{E}_2(s^{H_1} \mid H_1 < \infty)\mathbb{E}_2(s^{\widetilde{H}_0} \mid H_1 < \infty)\mathbb{P}_2(H_1 < \infty)$$
$$= \mathbb{E}_2(s^{H_1} 1_{H_1 < \infty})\mathbb{E}_2(s^{\widetilde{H}_0} \mid H_1 < \infty)$$
$$= \mathbb{E}_2(s^{H_1})^2 = \phi(s)^2.$$

Then, by the Markov property at time 1, conditional on $X_1 = 2$, we have $H_0 = 1 + \overline{H}_0$, where $\overline{H}_0$, the time taken after time 1 to get to 0, has the same distribution as H_0 does under $\mathbb{P}_2$. So
$$\phi(s) = \mathbb{E}_1(s^{H_0}) = p\mathbb{E}_1(s^{H_0} \mid X_1 = 2) + q\mathbb{E}_1(s^{H_0} \mid X_1 = 0)$$
$$= p\mathbb{E}_1(s^{1+\overline{H}_0} \mid X_1 = 2) + q\mathbb{E}_1(s \mid X_1 = 0)$$
$$= ps\mathbb{E}_2(s^{H_0}) + qs$$
$$= ps\phi(s)^2 + qs.$$

Thus $\phi = \phi(s)$ satisfies
$$ps\phi^2 - \phi + qs = 0 \tag{1.5}$$
and
$$\phi = (1 \pm \sqrt{1 - 4pqs^2})/2ps.$$

Since $\phi(0) \leq 1$ and ϕ is continuous we are forced to take the negative root at $s = 0$ and stick with it for all $0 \leq s < 1$.

To recover the distribution of H_0 we expand the square-root as a power series:
$$\phi(s) = \frac{1}{2ps}\left\{1 - \left(1 + \tfrac{1}{2}(-4pqs^2) + \tfrac{1}{2}(-\tfrac{1}{2})(-4pqs^2)^2/2! + \ldots\right)\right\}$$
$$= qs + pq^2s^3 + \ldots$$
$$= s\mathbb{P}_1(H_0 = 1) + s^2\mathbb{P}_1(H_0 = 2) + s^3\mathbb{P}_1(H_0 = 3) + \ldots.$$

The first few probabilities $\mathbb{P}_1(H_0 = 1), \mathbb{P}_1(H_0 = 2), \ldots$ are readily checked from first principles.

On letting $s \uparrow 1$ we have $\phi(s) \to \mathbb{P}_1(H_0 < \infty)$, so

$$\mathbb{P}_1(H_0 < \infty) = \frac{1 - \sqrt{1 - 4pq}}{2p} = \begin{cases} 1 & \text{if } p \leq q \\ q/p & \text{if } p > q. \end{cases}$$

(Remember that $q = 1 - p$, so

$$\sqrt{1 - 4pq} = \sqrt{1 - 4p + 4p^2} = |1 - 2p| = |2q - 1|.)$$

We can also find the mean hitting time using

$$\mathbb{E}_1(H_0) = \lim_{s \uparrow 1} \phi'(s).$$

It is only worth considering the case $p \leq q$, where the mean hitting time has a chance of being finite. Differentiate (1.5) to obtain

$$2ps\phi\phi' + p\phi^2 - \phi' + q = 0$$

so

$$\phi'(s) = (p\phi(s)^2 + q)/(1 - 2ps\phi(s)) \to 1/(1 - 2p) = 1/(q - p) \quad \text{as } s \uparrow 1.$$

See Example 5.1.1 for a connection with branching processes.

Example 1.4.4

We now consider an application of the strong Markov property to a Markov chain $(X_n)_{n \geq 0}$ observed only at certain times. In the first instance suppose that J is some subset of the state-space I and that we observe the chain only when it takes values in J. The resulting process $(Y_m)_{m \geq 0}$ may be obtained formally by setting $Y_m = X_{T_m}$, where

$$T_0 = \inf\{n \geq 0 : X_n \in J\}$$

and, for $m = 0, 1, 2, \ldots$

$$T_{m+1} = \inf\{n > T_m : X_n \in J\}.$$

Let us assume that $\mathbb{P}(T_m < \infty) = 1$ for all m. For each m we can check easily that T_m, the time of the mth visit to J, is a stopping time. So the strong Markov property applies to show, for $i_1, \ldots, i_{m+1} \in J$, that

$$\mathbb{P}(Y_{m+1} = i_{m+1} \mid Y_0 = i_1, \ldots, Y_m = i_m)$$
$$= \mathbb{P}(X_{T_{m+1}} = i_{m+1} \mid X_{T_0} = i_1, \ldots, X_{T_m} = i_m)$$
$$= \mathbb{P}_{i_m}(X_{T_1} = i_{m+1}) = \overline{p}_{i_m i_{m+1}}$$

where, for $i, j \in J$
$$\overline{p}_{ij} = h_i^j$$
and where, for $j \in J$, the vector $(h_i^j : i \in I)$ is the minimal non-negative solution to
$$h_i^j = p_{ij} + \sum_{k \notin J} p_{ik} h_k^j. \tag{1.6}$$

Thus $(Y_m)_{m \geq 0}$ is a Markov chain on J with transition matrix $\overline{P}$.

A second example of a similar type arises if we observe the original chain $(X_n)_{n \geq 0}$ only when it moves. The resulting process $(Z_m)_{m \geq 0}$ is given by $Z_m = X_{S_m}$ where $S_0 = 0$ and for $m = 0, 1, 2, \ldots$
$$S_{m+1} = \inf\{n \geq S_m : X_n \neq X_{S_m}\}.$$

Let us assume there are no absorbing states. Again the random times S_m for $m \geq 0$ are stopping times and, by the strong Markov property
$$\mathbb{P}(Z_{m+1} = i_{m+1} \mid Z_0 = i_1, \ldots, Z_m = i_m)$$
$$= \mathbb{P}(X_{S_{m+1}} = i_{m+1} \mid X_{S_0} = i_1, \ldots, X_{S_m} = i_m)$$
$$= \mathbb{P}_{i_m}(X_{S_1} = i_{m+1}) = \widetilde{p}_{i_m i_{m+1}}$$

where $\widetilde{p}_{ii} = 0$ and, for $i \neq j$
$$\widetilde{p}_{ij} = p_{ij} / \sum_{k \neq i} p_{ik}.$$

Thus $(Z_m)_{m \geq 0}$ is a Markov chain on I with transition matrix $\widetilde{P}$.

Exercises

1.4.1 Let $Y_1, Y_2, \ldots$ be independent identically distributed random variables with
$\mathbb{P}(Y_1 = 1) = \mathbb{P}(Y_1 = -1) = 1/2$ and set $X_0 = 1$, $X_n = X_0 + Y_1 + \ldots + Y_n$ for $n \geq 1$. Define
$$H_0 = \inf\{n \geq 0 : X_n = 0\}.$$
Find the probability generating function $\phi(s) = \mathbb{E}(s^{H_0})$.

Suppose the distribution of $Y_1, Y_2, \ldots$ is changed to $\mathbb{P}(Y_1 = 2) = 1/2$, $\mathbb{P}(Y_1 = -1) = 1/2$. Show that ϕ now satisfies
$$s\phi^3 - 2\phi + s = 0.$$

1.4.2 Deduce carefully from Theorem 1.3.2 the claim made at (1.6).

1.5 Recurrence and transience

Let $(X_n)_{n \geq 0}$ be a Markov chain with transition matrix P. We say that a state i is *recurrent* if

$$\mathbb{P}_i(X_n = i \text{ for infinitely many } n) = 1.$$

We say that i is *transient* if

$$\mathbb{P}_i(X_n = i \text{ for infinitely many } n) = 0.$$

Thus a recurrent state is one to which you keep coming back and a transient state is one which you eventually leave for ever. We shall show that every state is either recurrent or transient.

Recall that the *first passage time* to state i is the random variable T_i defined by

$$T_i(\omega) = \inf\{n \geq 1 : X_n(\omega) = i\}$$

where $\inf \emptyset = \infty$. We now define inductively the *rth passage time* $T_i^{(r)}$ to state i by

$$T_i^{(0)}(\omega) = 0, \quad T_i^{(1)}(\omega) = T_i(\omega)$$

and, for $r = 0, 1, 2, \ldots$,

$$T_i^{(r+1)}(\omega) = \inf\{n \geq T_i^{(r)}(\omega) + 1 : X_n(\omega) = i\}.$$

The length of the rth excursion to i is then

$$S_i^{(r)} = \begin{cases} T_i^{(r)} - T_i^{(r-1)} & \text{if } T_i^{(r-1)} < \infty \\ 0 & \text{otherwise.} \end{cases}$$

The following diagram illustrates these definitions:

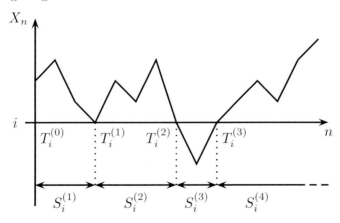

Our analysis of recurrence and transience will rest on finding the joint distribution of these excursion lengths.

Lemma 1.5.1. *For $r = 2, 3, \ldots$, conditional on $T_i^{(r-1)} < \infty$, $S_i^{(r)}$ is independent of $\{X_m : m \leq T_i^{(r-1)}\}$ and*

$$\mathbb{P}(S_i^{(r)} = n \mid T_i^{(r-1)} < \infty) = \mathbb{P}_i(T_i = n).$$

Proof. Apply the strong Markov property at the stopping time $T = T_i^{(r-1)}$. It is automatic that $X_T = i$ on $T < \infty$. So, conditional on $T < \infty$, $(X_{T+n})_{n \geq 0}$ is Markov(δ_i, P) and independent of $X_0, X_1, \ldots, X_T$. But

$$S_i^{(r)} = \inf\{n \geq 1 : X_{T+n} = i\},$$

so $S_i^{(r)}$ is the first passage time of $(X_{T+n})_{n \geq 0}$ to state i. □

Recall that the indicator function $1_{\{X_1 = j\}}$ is the random variable equal to 1 if $X_1 = j$ and 0 otherwise. Let us introduce the *number of visits* V_i to i, which may be written in terms of indicator functions as

$$V_i = \sum_{n=0}^{\infty} 1_{\{X_n = i\}}$$

and note that

$$\mathbb{E}_i(V_i) = \mathbb{E}_i \sum_{n=0}^{\infty} 1_{\{X_n = i\}} = \sum_{n=0}^{\infty} \mathbb{E}_i(1_{\{X_n = i\}}) = \sum_{n=0}^{\infty} \mathbb{P}_i(X_n = i) = \sum_{n=0}^{\infty} p_{ii}^{(n)}.$$

Also, we can compute the distribution of V_i under $\mathbb{P}_i$ in terms of the *return probability*

$$f_i = \mathbb{P}_i(T_i < \infty).$$

Lemma 1.5.2. *For $r = 0, 1, 2, \ldots$, we have $\mathbb{P}_i(V_i > r) = f_i^r$.*

Proof. Observe that if $X_0 = i$ then $\{V_i > r\} = \{T_i^{(r)} < \infty\}$. When $r = 0$ the result is true. Suppose inductively that it is true for r, then

$$\mathbb{P}_i(V_i > r+1) = \mathbb{P}_i(T_i^{(r+1)} < \infty)$$
$$= \mathbb{P}_i(T_i^{(r)} < \infty \text{ and } S_i^{(r+1)} < \infty)$$
$$= \mathbb{P}_i(S_i^{(r+1)} < \infty \mid T_i^{(r)} < \infty)\mathbb{P}_i(T_i^{(r)} < \infty)$$
$$= f_i f_i^r = f_i^{r+1}$$

by Lemma 1.5.1, so by induction the result is true for all r. □

Recall that one can compute the expectation of a non-negative integer-valued random variable as follows:

$$\sum_{r=0}^{\infty} \mathbb{P}(V > r) = \sum_{r=0}^{\infty} \sum_{v=r+1}^{\infty} \mathbb{P}(V = v)$$
$$= \sum_{v=1}^{\infty} \sum_{r=0}^{v-1} \mathbb{P}(V = v) = \sum_{v=1}^{\infty} v \mathbb{P}(V = v) = \mathbb{E}(V).$$

The next theorem is the means by which we establish recurrence or transience for a given state. Note that it provides two criteria for this, one in terms of the return probability, the other in terms of the n-step transition probabilities. Both are useful.

Theorem 1.5.3. *The following dichotomy holds:*
  (i) *if* $\mathbb{P}_i(T_i < \infty) = 1$, *then i is recurrent and* $\sum_{n=0}^{\infty} p_{ii}^{(n)} = \infty$;
  (ii) *if* $\mathbb{P}_i(T_i < \infty) < 1$, *then i is transient and* $\sum_{n=0}^{\infty} p_{ii}^{(n)} < \infty$.
In particular, every state is either transient or recurrent.

Proof. If $\mathbb{P}_i(T_i < \infty) = 1$, then, by Lemma 1.5.2,

$$\mathbb{P}_i(V_i = \infty) = \lim_{r \to \infty} \mathbb{P}_i(V_i > r) = 1$$

so i is recurrent and

$$\sum_{n=0}^{\infty} p_{ii}^{(n)} = \mathbb{E}_i(V_i) = \infty.$$

On the other hand, if $f_i = \mathbb{P}_i(T_i < \infty) < 1$, then by Lemma 1.5.2

$$\sum_{n=0}^{\infty} p_{ii}^{(n)} = \mathbb{E}_i(V_i) = \sum_{r=0}^{\infty} \mathbb{P}_i(V_i > r) = \sum_{r=0}^{\infty} f_i^r = \frac{1}{1 - f_i} < \infty$$

so $\mathbb{P}_i(V_i = \infty) = 0$ and i is transient. □

From this theorem we can go on to solve completely the problem of recurrence or transience for Markov chains with finite state-space. Some cases of infinite state-space are dealt with in the following chapter. First we show that recurrence and transience are *class properties*.

Theorem 1.5.4. *Let C be a communicating class. Then either all states in C are transient or all are recurrent.*

Proof. Take any pair of states $i, j \in C$ and suppose that i is transient. There exist $n, m \geq 0$ with $p_{ij}^{(n)} > 0$ and $p_{ji}^{(m)} > 0$, and, for all $r \geq 0$

$$p_{ii}^{(n+r+m)} \geq p_{ij}^{(n)} p_{jj}^{(r)} p_{ji}^{(m)}$$

so

$$\sum_{r=0}^{\infty} p_{jj}^{(r)} \leq \frac{1}{p_{ij}^{(n)} p_{ji}^{(m)}} \sum_{r=0}^{\infty} p_{ii}^{(n+r+m)} < \infty$$

by Theorem 1.5.3. Hence j is also transient by Theorem 1.5.3. □

In the light of this theorem it is natural to speak of a recurrent or transient class.

Theorem 1.5.5. *Every recurrent class is closed.*

Proof. Let C be a class which is not closed. Then there exist $i \in C$, $j \notin C$ and $m \geq 1$ with

$$\mathbb{P}_i(X_m = j) > 0.$$

Since we have

$$\mathbb{P}_i(\{X_m = j\} \cap \{X_n = i \text{ for infinitely many } n\}) = 0$$

this implies that

$$\mathbb{P}_i(X_n = i \text{ for infinitely many } n) < 1$$

so i is not recurrent, and so neither is C. □

Theorem 1.5.6. *Every finite closed class is recurrent.*

Proof. Suppose C is closed and finite and that $(X_n)_{n \geq 0}$ starts in C. Then for some $i \in C$ we have

$$0 < \mathbb{P}(X_n = i \text{ for infinitely many } n)$$
$$= \mathbb{P}(X_n = i \text{ for some } n) \mathbb{P}_i(X_n = i \text{ for infinitely many } n)$$

by the strong Markov property. This shows that i is not transient, so C is recurrent by Theorems 1.5.3 and 1.5.4. □

It is easy to spot closed classes, so the transience or recurrence of finite classes is easy to determine. For example, the only recurrent class in Example 1.2.2 is $\{5, 6\}$, the others being transient. On the other hand, infinite closed classes may be transient: see Examples 1.3.3 and 1.6.3.

We shall need the following result in Section 1.8. Remember that irreducibility means that the chain can get from any state to any other, with positive probability.

Theorem 1.5.7. *Suppose P is irreducible and recurrent. Then for all $j \in I$ we have $\mathbb{P}(T_j < \infty) = 1$.*

Proof. By the Markov property we have
$$\mathbb{P}(T_j < \infty) = \sum_{i \in I} \mathbb{P}(X_0 = i)\mathbb{P}_i(T_j < \infty)$$
so it suffices to show $\mathbb{P}_i(T_j < \infty) = 1$ for all $i \in I$. Choose m with $p_{ji}^{(m)} > 0$. By Theorem 1.5.3, we have
$$\begin{aligned}
1 &= \mathbb{P}_j(X_n = j \text{ for infinitely many } n) \\
&= \mathbb{P}_j(X_n = j \text{ for some } n \geq m+1) \\
&= \sum_{k \in I} \mathbb{P}_j(X_n = j \text{ for some } n \geq m+1 \mid X_m = k)\mathbb{P}_j(X_m = k) \\
&= \sum_{k \in I} \mathbb{P}_k(T_j < \infty)p_{jk}^{(m)}
\end{aligned}$$
where the final equality uses the Markov property. But $\sum_{k \in I} p_{jk}^{(m)} = 1$ so we must have $\mathbb{P}_i(T_j < \infty) = 1$. $\square$

Exercises

1.5.1 In Exercise 1.2.1, which states are recurrent and which are transient?

1.5.2 Show that, for the Markov chain $(X_n)_{n \geq 0}$ in Exercise 1.3.4 we have
$$\mathbb{P}(X_n \to \infty \text{ as } n \to \infty) = 1.$$
Suppose, instead, the transition probabilities satisfy
$$p_{i,i+1} = \left(\frac{i+1}{i}\right)^\alpha p_{i,i-1}.$$
For each $\alpha \in (0, \infty)$ find the value of $\mathbb{P}(X_n \to \infty \text{ as } n \to \infty)$.

1.5.3 (First passage decomposition). Denote by T_j the first passage time to state j and set
$$f_{ij}^{(n)} = \mathbb{P}_i(T_j = n).$$
Justify the identity
$$p_{ij}^{(n)} = \sum_{k=1}^n f_{ij}^{(k)} p_{jj}^{(n-k)} \qquad \text{for } n \geq 1$$

1.6 Recurrence and transience of random walks

and deduce that
$$P_{ij}(s) = \delta_{ij} + F_{ij}(s)P_{jj}(s)$$
where
$$P_{ij}(s) = \sum_{n=0}^{\infty} p_{ij}^{(n)} s^n, \quad F_{ij}(s) = \sum_{n=0}^{\infty} f_{ij}^{(n)} s^n.$$
Hence show that $\mathbb{P}_i(T_i < \infty) = 1$ if and only if
$$\sum_{n=0}^{\infty} p_{ii}^{(n)} = \infty$$
without using Theorem 1.5.3.

1.5.4 A random sequence of non-negative integers $(F_n)_{n\geq 0}$ is obtained by setting $F_0 = 0$ and $F_1 = 1$ and, once $F_0, \ldots, F_n$ are known, taking F_{n+1} to be either the sum or the difference of F_{n-1} and F_n, each with probability $1/2$. Is $(F_n)_{n\geq 0}$ a Markov chain?

By considering the Markov chain $X_n = (F_{n-1}, F_n)$, find the probability that $(F_n)_{n\geq 0}$ reaches 3 before first returning to 0.

Draw enough of the flow diagram for $(X_n)_{n\geq 0}$ to establish a general pattern. Hence, using the strong Markov property, show that the hitting probability for $(1,1)$, starting from $(1,2)$, is $(3-\sqrt{5})/2$.

Deduce that $(X_n)_{n\geq 0}$ is transient. Show that, moreover, with probability 1, $F_n \to \infty$ as $n \to \infty$.

1.6 Recurrence and transience of random walks

In the last section we showed that recurrence was a class property, that all recurrent classes were closed and that all finite closed classes were recurrent. So the only chains for which the question of recurrence remains interesting are irreducible with infinite state-space. Here we shall study some simple and fundamental examples of this type, making use of the following criterion for recurrence from Theorem 1.5.3: a state i is recurrent if and only if $\sum_{n=0}^{\infty} p_{ii}^{(n)} = \infty$.

Example 1.6.1 (Simple random walk on $\mathbb{Z}$)

The simple random walk on $\mathbb{Z}$ has diagram

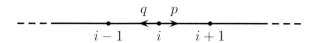

where $0 < p = 1 - q < 1$. Suppose we start at 0. It is clear that we cannot return to 0 after an odd number of steps, so $p_{00}^{(2n+1)} = 0$ for all n. Any given sequence of steps of length $2n$ from 0 to 0 occurs with probability $p^n q^n$, there being n steps up and n steps down, and the number of such sequences is the number of ways of choosing the n steps up from $2n$. Thus

$$p_{00}^{(2n)} = \binom{2n}{n} p^n q^n.$$

Stirling's formula provides a good approximation to $n!$ for large n: it is known that

$$n! \sim \sqrt{2\pi n}(n/e)^n \quad \text{as } n \to \infty$$

where $a_n \sim b_n$ means $a_n/b_n \to 1$. For a proof see W. Feller, *An Introduction to Probability Theory and its Applications, Vol I* (Wiley, New York, 3rd edition, 1968). At the end of this chapter we reproduce the argument used by Feller to show that

$$n! \sim A\sqrt{n}(n/e)^n \quad \text{as } n \to \infty$$

for some $A \in [1, \infty)$. The additional work needed to show $A = \sqrt{2\pi}$ is omitted, as this fact is unnecessary to our applications.

For the n-step transition probabilities we obtain

$$p_{00}^{(2n)} = \frac{(2n)!}{(n!)^2}(pq)^n \sim \frac{(4pq)^n}{A\sqrt{n/2}} \quad \text{as } n \to \infty.$$

In the symmetric case $p = q = 1/2$, so $4pq = 1$; then for some N and all $n \geq N$ we have

$$p_{00}^{(2n)} \geq \frac{1}{2A\sqrt{n}}$$

so

$$\sum_{n=N}^{\infty} p_{00}^{(2n)} \geq \frac{1}{2A} \sum_{n=N}^{\infty} \frac{1}{\sqrt{n}} = \infty$$

which shows that the random walk is recurrent. On the other hand, if $p \neq q$ then $4pq = r < 1$, so by a similar argument, for some N

$$\sum_{n=N}^{\infty} p_{00}^{(n)} \leq \frac{1}{A} \sum_{n=N}^{\infty} r^n < \infty$$

showing that the random walk is transient.

1.6 Recurrence and transience of random walks

Example 1.6.2 (Simple symmetric random walk on $\mathbb{Z}^2$)

The simple symmetric random walk on $\mathbb{Z}^2$ has diagram

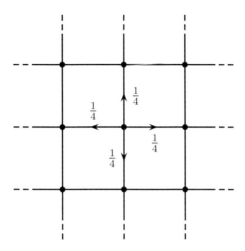

and transition probabilities

$$p_{ij} = \begin{cases} 1/4 & \text{if } |i-j| = 1 \\ 0 & \text{otherwise.} \end{cases}$$

Suppose we start at 0. Let us call the walk X_n and write X_n^+ and X_n^- for the orthogonal projections of X_n on the diagonal lines $y = \pm x$:

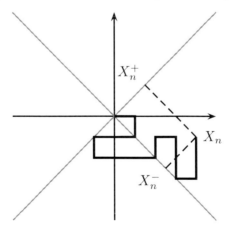

Then X_n^+ and X_n^- are independent simple symmetric random walks on $2^{-1/2}\mathbb{Z}$ and $X_n = 0$ if and only if $X_n^+ = 0 = X_n^-$. This makes it clear that for X_n we have

$$p_{00}^{(2n)} = \left(\binom{2n}{n}\left(\frac{1}{2}\right)^{2n}\right)^2 \sim \frac{2}{A^2 n} \qquad \text{as } n \to \infty$$

by Stirling's formula. Then $\sum_{n=1}^{\infty} p_{00}^{(n)} = \infty$ by comparison with $\sum_{n=1}^{\infty} 1/n$ and the walk is recurrent.

Example 1.6.3 (Simple symmetric random walk on $\mathbb{Z}^3$)

The transition probabilities of the simple symmetric random walk on $\mathbb{Z}^3$ are given by
$$p_{ij} = \begin{cases} 1/6 & \text{if } |i-j| = 1 \\ 0 & \text{otherwise.} \end{cases}$$

Thus the chain jumps to each of its nearest neighbours with equal probability. Suppose we start at 0. We can only return to 0 after an even number $2n$ of steps. Of these $2n$ steps there must be i up, i down, j north, j south, k east and k west for some $i, j, k \geq 0$, with $i + j + k = n$. By counting the ways in which this can be done, we obtain

$$p_{00}^{(2n)} = \sum_{\substack{i,j,k \geq 0 \\ i+j+k=n}} \frac{(2n)!}{(i!j!k!)^2} \left(\frac{1}{6}\right)^{2n} = \binom{2n}{n} \left(\frac{1}{2}\right)^{2n} \sum_{\substack{i,j,k \geq 0 \\ i+j+k=n}} \binom{n}{i\ j\ k}^2 \left(\frac{1}{3}\right)^{2n}.$$

Now
$$\sum_{\substack{i,j,k \geq 0 \\ i+j+k=n}} \binom{n}{i\ j\ k} \left(\frac{1}{3}\right)^n = 1$$

the left-hand side being the total probability of all the ways of placing n balls randomly into three boxes. For the case where $n = 3m$, we have

$$\binom{n}{i\ j\ k} = \frac{n!}{i!j!k!} \leq \binom{n}{m\ m\ m}$$

for all i, j, k, so

$$p_{00}^{(2n)} \leq \binom{2n}{n} \left(\frac{1}{2}\right)^{2n} \binom{n}{m\ m\ m} \left(\frac{1}{3}\right)^n \sim \frac{1}{2A^3} \left(\frac{6}{n}\right)^{3/2} \quad \text{as } n \to \infty$$

by Stirling's formula. Hence, $\sum_{m=0}^{\infty} p_{00}^{(6m)} < \infty$ by comparison with $\sum_{n=0}^{\infty} n^{-3/2}$. But $p_{00}^{(6m)} \geq (1/6)^2 p_{00}^{(6m-2)}$ and $p_{00}^{(6m)} \geq (1/6)^4 p_{00}^{(6m-4)}$ for all m so we must have

$$\sum_{n=0}^{\infty} p_{00}^{(n)} < \infty$$

and the walk is transient.

Exercises

1.6.1 The rooted binary tree is an infinite graph T with one distinguished vertex R from which comes a single edge; at every other vertex there are three edges and there are no closed loops. The random walk on T jumps from a vertex along each available edge with equal probability. Show that the random walk is transient.

1.6.2 Show that the simple symmetric random walk in $\mathbb{Z}^4$ is transient.

1.7 Invariant distributions

Many of the long-time properties of Markov chains are connected with the notion of an invariant distribution or measure. Remember that a measure λ is any row vector $(\lambda_i : i \in I)$ with non-negative entries. We say λ is *invariant* if
$$\lambda P = \lambda.$$

The terms *equilibrium* and *stationary* are also used to mean the same. The first result explains the term stationary.

Theorem 1.7.1. *Let $(X_n)_{n \geq 0}$ be Markov(λ, P) and suppose that λ is invariant for P. Then $(X_{m+n})_{n \geq 0}$ is also Markov(λ, P).*

Proof. By Theorem 1.1.3, $\mathbb{P}(X_m = i) = (\lambda P^m)_i = \lambda_i$ for all i and, clearly, conditional on $X_{m+n} = i$, X_{m+n+1} is independent of $X_m, X_{m+1}, \ldots, X_{m+n}$ and has distribution $(p_{ij} : j \in I)$. □

The next result explains the term equilibrium.

Theorem 1.7.2. *Let I be finite. Suppose for some $i \in I$ that*
$$p_{ij}^{(n)} \to \pi_j \quad \text{as} \quad n \to \infty \quad \text{for all } j \in I.$$

Then $\pi = (\pi_j : j \in I)$ is an invariant distribution.

Proof. We have
$$\sum_{j \in I} \pi_j = \sum_{j \in I} \lim_{n \to \infty} p_{ij}^{(n)} = \lim_{n \to \infty} \sum_{j \in I} p_{ij}^{(n)} = 1$$

and
$$\pi_j = \lim_{n \to \infty} p_{ij}^{(n)} = \lim_{n \to \infty} \sum_{k \in I} p_{ik}^{(n)} p_{kj} = \sum_{k \in I} \lim_{n \to \infty} p_{ik}^{(n)} p_{kj} = \sum_{k \in I} \pi_k p_{kj}$$

where we have used finiteness of I to justify interchange of summation and limit operations. Hence π is an invariant distribution. □

Notice that for any of the random walks discussed in Section 1.6 we have $p_{ij}^{(n)} \to 0$ as $n \to \infty$ for all $i, j \in I$. The limit is certainly invariant, but it is not a distribution!

Theorem 1.7.2 is not a very useful result but it serves to indicate a relationship between invariant distributions and n-step transition probabilities. In Theorem 1.8.3 we shall prove a sort of converse, which is much more useful.

Example 1.7.3

Consider the two-state Markov chain with transition matrix

$$P = \begin{pmatrix} 1-\alpha & \alpha \\ \beta & 1-\beta \end{pmatrix}.$$

Ignore the trivial cases $\alpha = \beta = 0$ and $\alpha = \beta = 1$. Then, by Example 1.1.4

$$P^n \to \begin{pmatrix} \beta/(\alpha+\beta) & \alpha/(\alpha+\beta) \\ \beta/(\alpha+\beta) & \alpha/(\alpha+\beta) \end{pmatrix} \quad \text{as } n \to \infty,$$

so, by Theorem 1.7.2, the distribution $(\beta/(\alpha+\beta), \alpha/(\alpha+\beta))$ must be invariant. There are of course easier ways to discover this.

Example 1.7.4

Consider the Markov chain $(X_n)_{n \geq 0}$ with diagram

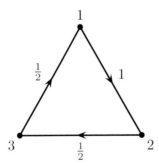

To find an invariant distribution we write down the components of the vector equation $\pi P = \pi$

$$\pi_1 = \tfrac{1}{2}\pi_3$$
$$\pi_2 = \pi_1 + \tfrac{1}{2}\pi_2$$
$$\pi_3 = \tfrac{1}{2}\pi_2 + \tfrac{1}{2}\pi_3.$$

In terms of the chain, the right-hand sides give the probabilities for X_1, when X_0 has distribution π, and the equations require X_1 also to have distribution π. The equations are homogeneous so one of them is redundant, and another equation is required to fix π uniquely. That equation is

$$\pi_1 + \pi_2 + \pi_3 = 1$$

and we find that $\pi = (1/5, 2/5, 2/5)$.

According to Example 1.1.6

$$p_{11}^{(n)} \to 1/5 \quad \text{as } n \to \infty$$

so this confirms Theorem 1.7.2. Alternatively, knowing that $p_{11}^{(n)}$ had the form

$$p_{11}^{(n)} = a + \left(\frac{1}{2}\right)^n \left(b \cos \frac{n\pi}{2} + c \sin \frac{n\pi}{2}\right)$$

we could have used Theorem 1.7.2 and knowledge of π_1 to identify $a = 1/5$, instead of working out $p_{11}^{(2)}$ in Example 1.1.6.

In the next two results we shall show that every irreducible and recurrent stochastic matrix P has an essentially unique positive invariant measure. The proofs rely heavily on the probabilistic interpretation so it is worth noting at the outset that, for a finite state-space I, the existence of an invariant row vector is a simple piece of linear algebra: the row sums of P are all 1, so the column vector of ones is an eigenvector with eigenvalue 1, so P must have a row eigenvector with eigenvalue 1.

For a fixed state k, consider for each i the *expected time spent in i between visits to k*:

$$\gamma_i^k = \mathbb{E}_k \sum_{n=0}^{T_k - 1} 1_{\{X_n = i\}}.$$

Here the sum of indicator functions serves to count the number of times n at which $X_n = i$ before the first passage time T_k.

Theorem 1.7.5. *Let P be irreducible and recurrent. Then*
  (i) $\gamma_k^k = 1$;
  (ii) $\gamma^k = (\gamma_i^k : i \in I)$ *satisfies* $\gamma^k P = \gamma^k$;
  (iii) $0 < \gamma_i^k < \infty$ *for all $i \in I$.*

Proof. (i) This is obvious. (ii) For $n = 1, 2, \ldots$ the event $\{n \leq T_k\}$ depends only on $X_0, X_1, \ldots, X_{n-1}$, so, by the Markov property at $n - 1$

$$\mathbb{P}_k(X_{n-1} = i, X_n = j \text{ and } n \leq T_k) = \mathbb{P}_k(X_{n-1} = i \text{ and } n \leq T_k) p_{ij}.$$

Since P is recurrent, under $\mathbb{P}_k$ we have $T_k < \infty$ and $X_0 = X_{T_k} = k$ with probability one. Therefore

$$\gamma_j^k = \mathbb{E}_k \sum_{n=1}^{T_k} 1_{\{X_n = j\}} = \mathbb{E}_k \sum_{n=1}^{\infty} 1_{\{X_n = j \text{ and } n \leq T_k\}}$$

$$= \sum_{n=1}^{\infty} \mathbb{P}_k(X_n = j \text{ and } n \leq T_k)$$

$$= \sum_{i \in I} \sum_{n=1}^{\infty} \mathbb{P}_k(X_{n-1} = i, X_n = j \text{ and } n \leq T_k)$$

$$= \sum_{i \in I} p_{ij} \sum_{n=1}^{\infty} \mathbb{P}_k(X_{n-1} = i \text{ and } n \leq T_k)$$

$$= \sum_{i \in I} p_{ij} \mathbb{E}_k \sum_{m=0}^{\infty} 1_{\{X_m = i \text{ and } m \leq T_k - 1\}}$$

$$= \sum_{i \in I} p_{ij} \mathbb{E}_k \sum_{m=0}^{T_k - 1} 1_{\{X_m = i\}} = \sum_{i \in I} \gamma_i^k p_{ij}.$$

(iii) Since P is irreducible, for each state i there exist $n, m \geq 0$ with $p_{ik}^{(n)}, p_{ki}^{(m)} > 0$. Then $\gamma_i^k \geq \gamma_k^k p_{ki}^{(m)} > 0$ and $\gamma_i^k p_{ik}^{(n)} \leq \gamma_k^k = 1$ by (i) and (ii). $\square$

Theorem 1.7.6. *Let P be irreducible and let λ be an invariant measure for P with $\lambda_k = 1$. Then $\lambda \geq \gamma^k$. If in addition P is recurrent, then $\lambda = \gamma^k$.*

Proof. For each $j \in I$ we have

$$\lambda_j = \sum_{i_1 \in I} \lambda_{i_1} p_{i_1 j} = \sum_{i_1 \neq k} \lambda_{i_1} p_{i_1 j} + p_{kj}$$

$$= \sum_{i_1, i_2 \neq k} \lambda_{i_2} p_{i_2 i_1} p_{i_1 j} + \left(p_{kj} + \sum_{i_1 \neq k} p_{k i_1} p_{i_1 j} \right)$$

$$\vdots$$

$$= \sum_{i_1, \ldots, i_n \neq k} \lambda_{i_n} p_{i_n i_{n-1}} \cdots p_{i_1 j}$$

$$+ \left(p_{kj} + \sum_{i_1 \neq k} p_{k i_1} p_{i_1 j} + \ldots + \sum_{i_1, \ldots, i_{n-1} \neq k} p_{k i_{n-1}} \cdots p_{i_2 i_1} p_{i_1 j} \right)$$

So for $j \neq k$ we obtain

$$\lambda_j \geq \mathbb{P}_k(X_1 = j \text{ and } T_k \geq 1) + \mathbb{P}_k(X_2 = j \text{ and } T_k \geq 2)$$
$$+ \ldots + \mathbb{P}_k(X_n = j \text{ and } T_k \geq n)$$
$$\to \gamma_j^k \quad \text{as } n \to \infty.$$

1.7 Invariant distributions

So $\lambda \geq \gamma^k$. If P is recurrent, then γ^k is invariant by Theorem 1.7.5, so $\mu = \lambda - \gamma^k$ is also invariant and $\mu \geq 0$. Since P is irreducible, given $i \in I$, we have $p_{ik}^{(n)} > 0$ for some n, and $0 = \mu_k = \sum_{j \in I} \mu_j p_{jk}^{(n)} \geq \mu_i p_{ik}^{(n)}$, so $\mu_i = 0$. $\square$

Recall that a state i is recurrent if

$$\mathbb{P}_i(X_n = i \text{ for infinitely many } n) = 1$$

and we showed in Theorem 1.5.3 that this is equivalent to

$$\mathbb{P}_i(T_i < \infty) = 1.$$

If in addition the *expected return time*

$$m_i = \mathbb{E}_i(T_i)$$

is finite, then we say i is *positive recurrent*. A recurrent state which fails to have this stronger property is called *null recurrent*.

Theorem 1.7.7. *Let P be irreducible. Then the following are equivalent:*
  (i) *every state is positive recurrent;*
  (ii) *some state i is positive recurrent;*
  (iii) *P has an invariant distribution, π say.*
  Moreover, when (iii) holds we have $m_i = 1/\pi_i$ for all i.

Proof. (i) $\Rightarrow$ (ii) This is obvious.
(ii) $\Rightarrow$ (iii) If i is positive recurrent, it is certainly recurrent, so P is recurrent. By Theorem 1.7.5, γ^i is then invariant. But

$$\sum_{j \in I} \gamma_j^i = m_i < \infty$$

so $\pi_j = \gamma_j^i / m_i$ defines an invariant distribution.
(iii) $\Rightarrow$ (i) Take any state k. Since P is irreducible and $\sum_{i \in I} \pi_i = 1$ we have $\pi_k = \sum_{i \in I} \pi_i p_{ik}^{(n)} > 0$ for some n. Set $\lambda_i = \pi_i / \pi_k$. Then λ is an invariant measure with $\lambda_k = 1$. So by Theorem 1.7.6, $\lambda \geq \gamma^k$. Hence

$$m_k = \sum_{i \in I} \gamma_i^k \leq \sum_{i \in I} \frac{\pi_i}{\pi_k} = \frac{1}{\pi_k} < \infty \tag{1.7}$$

and k is positive recurrent.

To complete the proof we return to the argument for (iii) ⇒ (i) armed with the knowledge that P is recurrent, so $\lambda = \gamma^k$ and the inequality (1.7) is in fact an equality. □

Example 1.7.8 (Simple symmetric random walk on $\mathbb{Z}$)

The simple symmetric random walk on $\mathbb{Z}$ is clearly irreducible and, by Example 1.6.1, it is also recurrent. Consider the measure

$$\pi_i = 1 \quad \text{for all } i.$$

Then

$$\pi_i = \tfrac{1}{2}\pi_{i-1} + \tfrac{1}{2}\pi_{i+1}$$

so π is invariant. Now Theorem 1.7.6 forces any invariant measure to be a scalar multiple of π. Since $\sum_{i \in \mathbb{Z}} \pi_i = \infty$, there can be no invariant distribution and the walk is therefore null recurrent, by Theorem 1.7.7.

Example 1.7.9

The existence of an invariant measure does not guarantee recurrence: consider, for example, the simple symmetric random walk on $\mathbb{Z}^3$, which is transient by Example 1.6.3, but has invariant measure π given by $\pi_i = 1$ for all i.

Example 1.7.10

Consider the asymmetric random walk on $\mathbb{Z}$ with transition probabilities $p_{i,i-1} = q < p = p_{i,i+1}$. In components the invariant measure equation $\pi P = \pi$ reads

$$\pi_i = \pi_{i-1}p + \pi_{i+1}q.$$

This is a recurrence relation for π with general solution

$$\pi_i = A + B(p/q)^i.$$

So, in this case, there is a two-parameter family of invariant measures – uniqueness up to scalar multiples does not hold.

Example 1.7.11

Consider a *success-run chain* on $\mathbb{Z}^+$, whose transition probabilities are given by

$$p_{i,i+1} = p_i, \quad p_{i0} = q_i = 1 - p_i.$$

Then the components of the invariant measure equation $\pi P = \pi$ read

$$\pi_0 = \sum_{i=0}^{\infty} q_i \pi_i,$$
$$\pi_i = p_{i-1} \pi_{i-1}, \quad \text{for } i \geq 1.$$

Suppose we choose p_i converging sufficiently rapidly to 1 so that

$$p = \prod_{i=0}^{\infty} p_i > 0.$$

Then for any invariant measure π we have

$$\pi_0 = \sum_{i=0}^{\infty} (1-p_i) p_{i-1} \ldots p_0 \pi_0 = (1-p) \pi_0.$$

This equation forces either $\pi_0 = 0$ or $\pi_0 = \infty$, so there is no non-zero invariant measure.

Exercises

1.7.1 Find all invariant distributions of the transition matrix in Exercise 1.2.1.

1.7.2 Gas molecules move about randomly in a box which is divided into two halves symmetrically by a partition. A hole is made in the partition. Suppose there are N molecules in the box. Show that the number of molecules on one side of the partition just after a molecule has passed through the hole evolves as a Markov chain. What are the transition probabilities? What is the invariant distribution of this chain?

1.7.3 A particle moves on the eight vertices of a cube in the following way: at each step the particle is equally likely to move to each of the three adjacent vertices, independently of its past motion. Let i be the initial vertex occupied by the particle, o the vertex opposite i. Calculate each of the following quantities:

(i) the expected number of steps until the particle returns to i;

(ii) the expected number of visits to o until the first return to i;

(iii) the expected number of steps until the first visit to o.

1.7.4 Let $(X_n)_{n\geq 0}$ be a simple random walk on $\mathbb{Z}$ with $p_{i,i-1} = q < p = p_{i,i+1}$. Find
$$\gamma_i^0 = \mathbb{E}_0\left(\sum_{n=0}^{T_0-1} 1_{\{X_n=i\}}\right)$$
and verify that
$$\gamma_i^0 = \inf_\lambda \lambda_i \quad \text{for all } i$$
where the infimum is taken over all invariant measures λ with $\lambda_0 = 1$. (Compare with Theorem 1.7.6 and Example 1.7.10.)

1.7.5 Let P be a stochastic matrix on a finite set I. Show that a distribution π is invariant for P if and only if $\pi(I - P + A) = a$, where $A = (a_{ij} : i, j \in I)$ with $a_{ij} = 1$ for all i and j, and $a = (a_i : i \in I)$ with $a_i = 1$ for all i. Deduce that if P is irreducible then $I - P + A$ is invertible. *Note that this enables one to compute the invariant distribution by any standard method of inverting a matrix.*

1.8 Convergence to equilibrium

We shall investigate the limiting behaviour of the n-step transition probabilities $p_{ij}^{(n)}$ as $n \to \infty$. As we saw in Theorem 1.7.2, if the state-space is finite and if for some i the limit exists for all j, then it must be an invariant distribution. But, as the following example shows, the limit does not always exist.

Example 1.8.1

Consider the two-state chain with transition matrix
$$P = \begin{pmatrix} 0 & 1 \\ 1 & 0 \end{pmatrix}.$$

Then $P^2 = I$, so $P^{2n} = I$ and $P^{2n+1} = P$ for all n. Thus $p_{ij}^{(n)}$ fails to converge for all i, j.

Let us call a state i *aperiodic* if $p_{ii}^{(n)} > 0$ for all sufficiently large n. We leave it as an exercise to show that i is aperiodic if and only if the set $\{n \geq 0 : p_{ii}^{(n)} > 0\}$ has no common divisor other than 1. This is also a consequence of Theorem 1.8.4. The behaviour of the chain in Example 1.8.1 is connected with its periodicity.

Lemma 1.8.2. Suppose P is irreducible and has an aperiodic state i. Then, for all states j and k, $p_{jk}^{(n)} > 0$ for all sufficiently large n. In particular, all states are aperiodic.

Proof. There exist $r, s \geq 0$ with $p_{ji}^{(r)}, p_{ik}^{(s)} > 0$. Then
$$p_{jk}^{(r+n+s)} \geq p_{ji}^{(r)} p_{ii}^{(n)} p_{ik}^{(s)} > 0$$
for all sufficiently large n. □

Here is the main result of this section. The method of proof, by coupling two Markov chains, is ingenious.

Theorem 1.8.3 (Convergence to equilibrium). Let P be irreducible and aperiodic, and suppose that P has an invariant distribution π. Let λ be any distribution. Suppose that $(X_n)_{n \geq 0}$ is Markov(λ, P). Then
$$\mathbb{P}(X_n = j) \to \pi_j \quad \text{as } n \to \infty \text{ for all } j.$$

In particular,
$$p_{ij}^{(n)} \to \pi_j \quad \text{as } n \to \infty \text{ for all } i, j.$$

Proof. We use a coupling argument. Let $(Y_n)_{n \geq 0}$ be Markov(π, P) and independent of $(X_n)_{n \geq 0}$. Fix a reference state b and set
$$T = \inf\{n \geq 1 : X_n = Y_n = b\}.$$

Step 1. We show $\mathbb{P}(T < \infty) = 1$. The process $W_n = (X_n, Y_n)$ is a Markov chain on $I \times I$ with transition probabilities
$$\widetilde{p}_{(i,k)(j,l)} = p_{ij} p_{kl}$$
and initial distribution
$$\mu_{(i,k)} = \lambda_i \pi_k.$$
Since P is aperiodic, for all states i, j, k, l we have
$$\widetilde{p}_{(i,k)(j,l)}^{(n)} = p_{ij}^{(n)} p_{kl}^{(n)} > 0$$
for all sufficiently large n; so $\widetilde{P}$ is irreducible. Also, $\widetilde{P}$ has an invariant distribution given by
$$\widetilde{\pi}_{(i,k)} = \pi_i \pi_k$$
so, by Theorem 1.7.7, $\widetilde{P}$ is positive recurrent. But T is the first passage time of W_n to (b, b) so $\mathbb{P}(T < \infty) = 1$, by Theorem 1.5.7.

Step 2. Set
$$Z_n = \begin{cases} X_n & \text{if } n < T \\ Y_n & \text{if } n \geq T. \end{cases}$$

The diagram below illustrates the idea. We show that $(Z_n)_{n \geq 0}$ is Markov(λ, P).

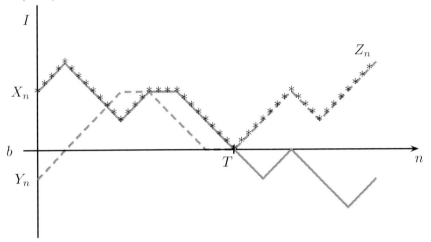

The strong Markov property applies to $(W_n)_{n \geq 0}$ at time T, so $(X_{T+n}, Y_{T+n})_{n \geq 0}$ is Markov$(\delta_{(b,b)}, \widetilde{P})$ and independent of (X_0, Y_0), $(X_1, Y_1), \ldots, (X_T, Y_T)$. By symmetry, we can replace the process $(X_{T+n}, Y_{T+n})_{n \geq 0}$ by $(Y_{T+n}, X_{T+n})_{n \geq 0}$ which is also Markov$(\delta_{(b,b)}, \widetilde{P})$ and remains independent of $(X_0, Y_0), (X_1, Y_1), \ldots, (X_T, Y_T)$. Hence $W'_n = (Z_n, Z'_n)$ is Markov$(\mu, \widetilde{P})$ where

$$Z'_n = \begin{cases} Y_n & \text{if } n < T \\ X_n & \text{if } n \geq T. \end{cases}$$

In particular, $(Z_n)_{n \geq 0}$ is Markov(λ, P).

Step 3. We have
$$\mathbb{P}(Z_n = j) = \mathbb{P}(X_n = j \text{ and } n < T) + \mathbb{P}(Y_n = j \text{ and } n \geq T)$$

so
$$\begin{aligned} |\mathbb{P}(X_n = j) - \pi_j| &= |\mathbb{P}(Z_n = j) - \mathbb{P}(Y_n = j)| \\ &= |\mathbb{P}(X_n = j \text{ and } n < T) - \mathbb{P}(Y_n = j \text{ and } n < T)| \\ &\leq \mathbb{P}(n < T) \end{aligned}$$

and $\mathbb{P}(n < T) \to 0$ as $n \to \infty$. $\square$

1.8 Convergence to equilibrium

To understand this proof one should see what goes wrong when P is not aperiodic. Consider the two-state chain of Example 1.8.1 which has $(1/2, 1/2)$ as its unique invariant distribution. We start $(X_n)_{n\geq 0}$ from 0 and $(Y_n)_{n\geq 0}$ with equal probability from 0 or 1. However, if $Y_0 = 1$, then, because of periodicity, $(X_n)_{n\geq 0}$ and $(Y_n)_{n\geq 0}$ will never meet, and the proof fails. We move on now to the cases that were excluded in the last theorem, where $(X_n)_{n\geq 0}$ is periodic or transient or null recurrent. The remainder of this section might be omitted on a first reading.

Theorem 1.8.4. *Let P be irreducible. There is an integer $d \geq 1$ and a partition*
$$I = C_0 \cup C_1 \cup \ldots \cup C_{d-1}$$
such that (setting $C_{nd+r} = C_r$)
  (i) $p_{ij}^{(n)} > 0$ *only if $i \in C_r$ and $j \in C_{r+n}$ for some r;*
  (ii) $p_{ij}^{(nd)} > 0$ *for all sufficiently large n, for all $i, j \in C_r$, for all r.*

Proof. Fix a state k and consider $S = \{n \geq 0 : p_{kk}^{(n)} > 0\}$. Choose $n_1, n_2 \in S$ with $n_1 < n_2$ and such that $d := n_2 - n_1$ is as small as possible. (Here and throughout we use the symbol $:=$ to mean 'defined to equal'.) Define for $r = 0, \ldots, d-1$
$$C_r = \{i \in I : p_{ki}^{(nd+r)} > 0 \text{ for some } n \geq 0\}.$$

Then $C_0 \cup \ldots \cup C_{d-1} = I$, by irreducibility. Moreover, if $p_{ki}^{(nd+r)} > 0$ and $p_{ki}^{(nd+s)} > 0$ for some $r, s \in \{0, 1, \ldots, d-1\}$, then, choosing $m \geq 0$ so that $p_{ik}^{(m)} > 0$, we have $p_{kk}^{(nd+r+m)} > 0$ and $p_{kk}^{(nd+s+m)} > 0$ so $r = s$ by minimality of d. Hence we have a partition.

To prove (i) suppose $p_{ij}^{(n)} > 0$ and $i \in C_r$. Choose m so that $p_{ki}^{(md+r)} > 0$, then $p_{kj}^{(md+r+n)} > 0$ so $j \in C_{r+n}$ as required. By taking $i = j = k$ we now see that d must divide every element of S, in particular n_1.

Now for $nd \geq n_1^2$, we can write $nd = qn_1 + r$ for integers $q \geq n_1$ and $0 \leq r \leq n_1 - 1$. Since d divides n_1 we then have $r = md$ for some integer m and then $nd = (q-m)n_1 + mn_2$. Hence
$$p_{kk}^{(nd)} \geq (p_{kk}^{(n_1)})^{q-m}(p_{kk}^{(n_2)})^m > 0$$
and hence $nd \in S$. To prove (ii) for $i, j \in C_r$ choose m_1 and m_2 so that $p_{ik}^{(m_1)} > 0$ and $p_{kj}^{(m_2)} > 0$, then
$$p_{ij}^{(m_1+nd+m_2)} \geq p_{ik}^{(m_1)} p_{kk}^{(nd)} p_{kj}^{(m_2)} > 0$$

whenever $nd \geq n_1^2$. Since $m_1 + m_2$ is then necessarily a multiple of d, we are done. □

We call d the *period* of P. The theorem just proved shows in particular for all $i \in I$ that d is the greatest common divisor of the set $\{n \geq 0 : p_{ii}^{(n)} > 0\}$. This is sometimes useful in identifying d.

Finally, here is a complete description of limiting behaviour for irreducible chains. This generalizes Theorem 1.8.3 in two respects since we require neither aperiodicity nor the existence of an invariant distribution. The argument we use for the null recurrent case was discovered recently by B. Fristedt and L. Gray.

Theorem 1.8.5. *Let P be irreducible of period d and let $C_0, C_1, \ldots, C_{d-1}$ be the partition obtained in Theorem 1.8.4. Let λ be a distribution with $\sum_{i \in C_0} \lambda_i = 1$. Suppose that $(X_n)_{n \geq 0}$ is Markov(λ, P). Then for $r = 0, 1, \ldots, d-1$ and $j \in C_r$ we have*

$$\mathbb{P}(X_{nd+r} = j) \to d/m_j \quad \text{as } n \to \infty$$

where m_j is the expected return time to j. In particular, for $i \in C_0$ and $j \in C_r$ we have

$$p_{ij}^{(nd+r)} \to d/m_j \quad \text{as } n \to \infty.$$

Proof

Step 1. We reduce to the aperiodic case. Set $\nu = \lambda P^r$, then by Theorem 1.8.4 we have

$$\sum_{i \in C_r} \nu_i = 1.$$

Set $Y_n = X_{nd+r}$, then $(Y_n)_{n \geq 0}$ is Markov(ν, P^d) and, by Theorem 1.8.4, P^d is irreducible and aperiodic on C_r. For $j \in C_r$ the expected return time of $(Y_n)_{n \geq 0}$ to j is m_j/d. So if the theorem holds in the aperiodic case, then

$$\mathbb{P}(X_{nd+r} = j) = \mathbb{P}(Y_n = j) \to d/m_j \quad \text{as } n \to \infty$$

so the theorem holds in general.

Step 2. Assume that P is aperiodic. If P is positive recurrent then $1/m_j = \pi_j$, where π is the unique invariant distribution, so the result follows from Theorem 1.8.3. Otherwise $m_j = \infty$ and we have to show that

$$\mathbb{P}(X_n = j) \to 0 \quad \text{as } n \to \infty.$$

If P is transient this is easy and we are left with the null recurrent case.

Step 3. Assume that P is aperiodic and null recurrent. Then
$$\sum_{k=0}^{\infty} \mathbb{P}_j(T_j > k) = \mathbb{E}_j(T_j) = \infty.$$

Given $\varepsilon > 0$ choose K so that
$$\sum_{k=0}^{K-1} \mathbb{P}_j(T_j > k) \geq \frac{2}{\varepsilon}.$$

Then, for $n \geq K - 1$
$$1 \geq \sum_{k=n-K+1}^{n} \mathbb{P}(X_k = j \text{ and } X_m \neq j \text{ for } m = k+1, \ldots, n)$$
$$= \sum_{k=n-K+1}^{n} \mathbb{P}(X_k = j)\mathbb{P}_j(T_j > n-k)$$
$$= \sum_{k=0}^{K-1} \mathbb{P}(X_{n-k} = j)\mathbb{P}_j(T_j > k)$$

so we must have $\mathbb{P}(X_{n-k} = j) \leq \varepsilon/2$ for some $k \in \{0, 1, \ldots, K-1\}$.

Return now to the coupling argument used in Theorem 1.8.3, only now let $(Y_n)_{n \geq 0}$ be Markov(μ, P), where μ is to be chosen later. Set $W_n = (X_n, Y_n)$. As before, aperiodicity of $(X_n)_{n \geq 0}$ ensures irreducibility of $(W_n)_{n \geq 0}$. If $(W_n)_{n \geq 0}$ is transient then, on taking $\mu = \lambda$, we obtain
$$\mathbb{P}(X_n = j)^2 = \mathbb{P}(W_n = (j,j)) \to 0$$

as required. Assume then that $(W_n)_{n \geq 0}$ is recurrent. Then, in the notation of Theorem 1.8.3, we have $\mathbb{P}(T < \infty) = 1$ and the coupling argument shows that
$$|\mathbb{P}(X_n = j) - \mathbb{P}(Y_n = j)| \to 0 \quad \text{as } n \to \infty.$$

We exploit this convergence by taking $\mu = \lambda P^k$ for $k = 1, \ldots, K-1$, so that $\mathbb{P}(Y_n = j) = \mathbb{P}(X_{n+k} = j)$. We can find N such that for $n \geq N$ and $k = 1, \ldots, K-1$,
$$|\mathbb{P}(X_n = j) - \mathbb{P}(X_{n+k} = j)| \leq \frac{\varepsilon}{2}.$$

But for any n we can find $k \in \{0, 1, \ldots, K-1\}$ such that $\mathbb{P}(X_{n+k} = j) \leq \varepsilon/2$. Hence, for $n \geq N$
$$\mathbb{P}(X_n = j) \leq \varepsilon.$$
Since $\varepsilon > 0$ was arbitrary, this shows that $\mathbb{P}(X_n = j) \to 0$ as $n \to \infty$, as required. $\square$

Exercises

1.8.1 Prove the claims (e), (f) and (g) made in example (v) of the Introduction.

1.8.2 Find the invariant distributions of the transition matrices in Exercise 1.1.7, parts (a), (b) and (c), and compare them with your answers there.

1.8.3 A fair die is thrown repeatedly. Let X_n denote the sum of the first n throws. Find
$$\lim_{n\to\infty} \mathbb{P}(X_n \text{ is a multiple of } 13)$$
quoting carefully any general theorems that you use.

1.8.4 Each morning a student takes one of the three books he owns from his shelf. The probability that he chooses book i is α_i, where $0 < \alpha_i < 1$ for $i = 1, 2, 3$, and choices on successive days are independent. In the evening he replaces the book at the left-hand end of the shelf. If p_n denotes the probability that on day n the student finds the books in the order 1,2,3, from left to right, show that, irrespective of the initial arrangement of the books, p_n converges as $n \to \infty$, and determine the limit.

1.8.5 (Renewal theorem). Let $Y_1, Y_2, \ldots$ be independent, identically distributed random variables with values in $\{1, 2, \ldots\}$. Suppose that the set of integers
$$\{n : \mathbb{P}(Y_1 = n) > 0\}$$
has greatest common divisor 1. Set $\mu = \mathbb{E}(Y_1)$. Show that the following process is a Markov chain:
$$X_n = \inf\{m \geq n : m = Y_1 + \ldots + Y_k \text{ for some } k \geq 0\} - n.$$
Determine
$$\lim_{n\to\infty} \mathbb{P}(X_n = 0)$$
and hence show that as $n \to \infty$
$$\mathbb{P}(n = Y_1 + \ldots + Y_k \text{ for some } k \geq 0) \to 1/\mu.$$

(*Think of $Y_1, Y_2, \ldots$ as light-bulb lifetimes. A bulb is replaced when it fails. Thus the limiting probability that a bulb is replaced at time n is $1/\mu$. Although this appears to be a very special case of convergence to equilibrium, one can actually recover the full result by applying the renewal theorem to the excursion lengths $S_i^{(1)}, S_i^{(2)}, \ldots$ from state i.*)

1.9 Time reversal

For Markov chains, the past and future are independent given the present. This property is symmetrical in time and suggests looking at Markov chains with time running backwards. On the other hand, convergence to equilibrium shows behaviour which is asymmetrical in time: a highly organised state such as a point mass decays to a disorganised one, the invariant distribution. This is an example of entropy increasing. It suggests that if we want complete time-symmetry we must begin in equilibrium. The next result shows that a Markov chain in equilibrium, run backwards, is again a Markov chain. The transition matrix may however be different.

Theorem 1.9.1. *Let P be irreducible and have an invariant distribution π. Suppose that $(X_n)_{0 \leq n \leq N}$ is Markov(π, P) and set $Y_n = X_{N-n}$. Then $(Y_n)_{0 \leq n \leq N}$ is Markov$(\pi, \widehat{P})$, where $\widehat{P} = (\widehat{p}_{ij})$ is given by*

$$\pi_j \widehat{p}_{ji} = \pi_i p_{ij} \quad \text{for all } i, j$$

and $\widehat{P}$ is also irreducible with invariant distribution π.

Proof. First we check that $\widehat{P}$ is a stochastic matrix:

$$\sum_{i \in I} \widehat{p}_{ji} = \frac{1}{\pi_j} \sum_{i \in I} \pi_i p_{ij} = 1$$

since π is invariant for P. Next we check that π is invariant for $\widehat{P}$:

$$\sum_{j \in I} \pi_j \widehat{p}_{ji} = \sum_{j \in I} \pi_i p_{ij} = \pi_i$$

since P is a stochastic matrix.

We have

$$\mathbb{P}(Y_0 = i_1, Y_1 = i_2, \ldots, Y_N = i_N)$$
$$= \mathbb{P}(X_0 = i_N, X_1 = i_{N-1}, \ldots, X_N = i_1)$$
$$= \pi_{i_N} p_{i_N i_{N-1}} \ldots p_{i_2 i_1} = \pi_{i_1} \widehat{p}_{i_1 i_2} \ldots \widehat{p}_{i_{N-1} i_N}$$

so, by Theorem 1.1.1, $(Y_n)_{0 \leq n \leq N}$ is Markov$(\pi, \widehat{P})$. Finally, since P is irreducible, for each pair of states i, j there is a chain of states $i_1 = i, i_2, \ldots, i_{n-1}, i_n = j$ with $p_{i_1 i_2} \ldots p_{i_{n-1} i_n} > 0$. Then
$$\widehat{p}_{i_n i_{n-1}} \ldots \widehat{p}_{i_2 i_1} = \pi_{i_1} p_{i_1 i_2} \ldots p_{i_{n-1} i_n} / \pi_{i_n} > 0$$
so $\widehat{P}$ is also irreducible. □

The chain $(Y_n)_{0 \leq n \leq N}$ is called the *time-reversal* of $(X_n)_{0 \leq n \leq N}$.

A stochastic matrix P and a measure λ are said to be in *detailed balance* if
$$\lambda_i p_{ij} = \lambda_j p_{ji} \quad \text{for all } i, j.$$
Though obvious, the following result is worth remembering because, when a solution λ to the detailed balance equations exists, it is often easier to find by the detailed balance equations than by the equation $\lambda = \lambda P$.

Lemma 1.9.2. *If P and λ are in detailed balance, then λ is invariant for P.*

Proof. We have $(\lambda P)_i = \sum_{j \in I} \lambda_j p_{ji} = \sum_{j \in I} \lambda_i p_{ij} = \lambda_i$. □

Let $(X_n)_{n \geq 0}$ be Markov(λ, P), with P irreducible. We say that $(X_n)_{n \geq 0}$ is *reversible* if, for all $N \geq 1$, $(X_{N-n})_{0 \leq n \leq N}$ is also Markov(λ, P).

Theorem 1.9.3. *Let P be an irreducible stochastic matrix and let λ be a distribution. Suppose that $(X_n)_{n \geq 0}$ is Markov(λ, P). Then the following are equivalent:*
  (a) *$(X_n)_{n \geq 0}$ is reversible;*
  (b) *P and λ are in detailed balance.*

Proof. Both (a) and (b) imply that λ is invariant for P. Then both (a) and (b) are equivalent to the statement that $\widehat{P} = P$ in Theorem 1.9.1. □

We begin a collection of examples with a chain which is not reversible.

Example 1.9.4

Consider the Markov chain with diagram:

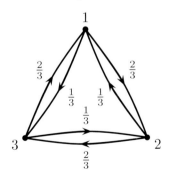

The transition matrix is

$$P = \begin{pmatrix} 0 & 2/3 & 1/3 \\ 1/3 & 0 & 2/3 \\ 2/3 & 1/3 & 0 \end{pmatrix}$$

and $\pi = (1/3, 1/3, 1/3)$ is invariant. Hence $\widehat{P} = P^T$, the transpose of P. But P is not symmetric, so $P \neq \widehat{P}$ and this chain is not reversible. A patient observer would see the chain move clockwise in the long run: under time-reversal the clock would run backwards!

Example 1.9.5

Consider the Markov chain with diagram:

where $0 < p = 1 - q < 1$. The non-zero detailed balance equations read

$$\lambda_i p_{i,i+1} = \lambda_{i+1} p_{i+1,i} \quad \text{for } i = 0, 1, \ldots, M-1.$$

So a solution is given by

$$\lambda = \big((p/q)^i : i = 0, 1, \ldots, M\big)$$

and this may be normalised to give a distribution in detailed balance with P. Hence this chain is reversible.

If p were much larger than q, one might argue that the chain would tend to move to the right and its time-reversal to the left. However, this ignores the fact that we reverse the chain *in equilibrium*, which in this case would be heavily concentrated near M. An observer would see the chain spending most of its time near M and making occasional brief forays to the left, which behaviour is symmetrical in time.

Example 1.9.6 (Random walk on a graph)

A *graph* G is a countable collection of states, usually called *vertices*, some of which are joined by *edges*, for example:

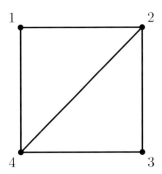

Thus a graph is a partially drawn Markov chain diagram. There is a natural way to complete the diagram which gives rise to the random walk on G. The *valency* v_i of vertex i is the number of edges at i. We have to assume that every vertex has finite valency. The random walk on G picks edges with equal probability:

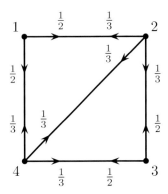

Thus the transition probabilities are given by

$$p_{ij} = \begin{cases} 1/v_i & \text{if } (i,j) \text{ is an edge} \\ 0 & \text{otherwise.} \end{cases}$$

We assume G is connected, so that P is irreducible. It is easy to see that P is in detailed balance with $v = (v_i : i \in G)$. So, if the total valency $\sigma = \sum_{i \in G} v_i$ is finite, then $\pi = v/\sigma$ is invariant and P is reversible.

Example 1.9.7 (Random chessboard knight)

A random knight makes each permissible move with equal probability. If it starts in a corner, how long on average will it take to return?

This is an example of a random walk on a graph: the vertices are the squares of the chessboard and the edges are the moves that the knight can take:

1.9 Time reversal

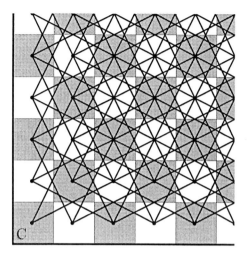

The diagram shows a part of the graph. We know by Theorem 1.7.7 and the preceding example that

$$\mathbb{E}_c(T_c) = 1/\pi_c = \sum_i (v_i/v_c)$$

so all we have to do is identify valencies. The four corner squares have valency 2, and the eight squares adjacent to the corners have valency 3. There are 20 squares of valency 4, 16 of valency 6, and the 16 central squares have valency 8. Hence

$$\mathbb{E}_c(T_c) = \frac{8 + 24 + 80 + 96 + 128}{2} = 168.$$

Alternatively, if you enjoy solving sets of 64 simultaneous linear equations, you might try finding π from $\pi P = \pi$, or calculating $\mathbb{E}_c(T_c)$ using Theorem 1.3.5!

Exercises

1.9.1 In each of the following cases determine whether the stochastic matrix P, which you may assume is irreducible, is reversible:

(a) $\begin{pmatrix} 1-p & p \\ q & 1-q \end{pmatrix}$;   (b) $\begin{pmatrix} 0 & p & 1-p \\ 1-p & 0 & p \\ p & 1-p & 0 \end{pmatrix}$;

(c) $I = \{0, 1, \ldots, N\}$ and $p_{ij} = 0$ if $|j - i| \geq 2$;

(d) $I = \{0, 1, 2, \dots\}$ and $p_{01} = 1$, $p_{i,i+1} = p$, $p_{i,i-1} = 1 - p$ for $i \geq 1$;

(e) $p_{ij} = p_{ji}$ for all $i, j \in S$.

1.9.2 Two particles X and Y perform independent random walks on the graph shown in the diagram. So, for example, a particle at A jumps to B, C or D with equal probability $1/3$.

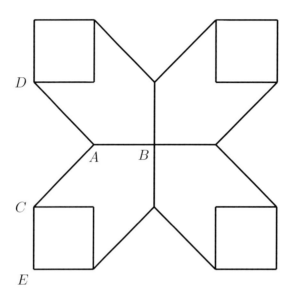

Find the probability that X and Y ever meet at a vertex in the following cases:

(a) X starts at A and Y starts at B;

(b) X starts at A and Y starts at E.

For $I = B, D$ let M_I denote the expected time, when both X and Y start at I, until they are once again both at I. Show that $9M_D = 16M_B$.

1.10 Ergodic theorem

Ergodic theorems concern the limiting behaviour of averages over time. We shall prove a theorem which identifies for Markov chains the long-run proportion of time spent in each state. An essential tool is the following ergodic theorem for independent random variables which is a version of the strong law of large numbers.

Theorem 1.10.1 (Strong law of large numbers). *Let $Y_1, Y_2, \dots$ be a sequence of independent, identically distributed, non-negative random*

variables with $\mathbb{E}(Y_1) = \mu$. Then
$$\mathbb{P}\left(\frac{Y_1 + \ldots + Y_n}{n} \to \mu \text{ as } n \to \infty\right) = 1.$$

Proof. A proof for the case $\mu < \infty$ may be found, for example, in *Probability with Martingales* by David Williams (Cambridge University Press, 1991). The case where $\mu = \infty$ is a simple deduction. Fix $N < \infty$ and set $Y_n^{(N)} = Y_n \wedge N$. Then
$$\frac{Y_1 + \ldots + Y_n}{n} \geq \frac{Y_1^{(N)} + \ldots + Y_n^{(N)}}{n} \to \mathbb{E}(Y_1 \wedge N) \qquad \text{as } n \to \infty$$
with probability one. As $N \uparrow \infty$ we have $\mathbb{E}(Y_1 \wedge N) \uparrow \mu$ by monotone convergence (see Section 6.4). So we must have, with probability 1
$$\frac{Y_1 + \ldots + Y_n}{n} \to \infty \qquad \text{as } n \to \infty. \qquad \square$$

We denote by $V_i(n)$ the *number of visits to i before n*:
$$V_i(n) = \sum_{k=0}^{n-1} 1_{\{X_k = i\}}.$$

Then $V_i(n)/n$ is the proportion of time before n spent in state i. The following result gives the long-run proportion of time spent by a Markov chain in each state.

Theorem 1.10.2 (Ergodic theorem). *Let P be irreducible and let λ be any distribution. If $(X_n)_{n \geq 0}$ is Markov(λ, P) then*
$$\mathbb{P}\left(\frac{V_i(n)}{n} \to \frac{1}{m_i} \text{ as } n \to \infty\right) = 1$$
where $m_i = \mathbb{E}_i(T_i)$ is the expected return time to state i. Moreover, in the positive recurrent case, for any bounded function $f : I \to \mathbb{R}$ we have
$$\mathbb{P}\left(\frac{1}{n} \sum_{k=0}^{n-1} f(X_k) \to \overline{f} \text{ as } n \to \infty\right) = 1$$
where
$$\overline{f} = \sum_{i \in I} \pi_i f_i$$
and where $(\pi_i : i \in I)$ is the unique invariant distribution.

Proof. If P is transient, then, with probability 1, the total number V_i of visits to i is finite, so

$$\frac{V_i(n)}{n} \leq \frac{V_i}{n} \to 0 = \frac{1}{m_i}.$$

Suppose then that P is recurrent and fix a state i. For $T = T_i$ we have $\mathbb{P}(T < \infty) = 1$ by Theorem 1.5.7 and $(X_{T+n})_{n \geq 0}$ is Markov(δ_i, P) and independent of $X_0, X_1, \ldots, X_T$ by the strong Markov property. The long-run proportion of time spent in i is the same for $(X_{T+n})_{n \geq 0}$ and $(X_n)_{n \geq 0}$, so it suffices to consider the case $\lambda = \delta_i$.

Write $S_i^{(r)}$ for the length of the rth excursion to i, as in Section 1.5. By Lemma 1.5.1, the non-negative random variables $S_i^{(1)}, S_i^{(2)}, \ldots$ are independent and identically distributed with $\mathbb{E}_i(S_i^{(r)}) = m_i$. Now

$$S_i^{(1)} + \ldots + S_i^{(V_i(n)-1)} \leq n - 1,$$

the left-hand side being the time of the last visit to i before n. Also

$$S_i^{(1)} + \ldots + S_i^{(V_i(n))} \geq n,$$

the left-hand side being the time of the first visit to i after $n - 1$. Hence

$$\frac{S_i^{(1)} + \ldots + S_i^{(V_i(n)-1)}}{V_i(n)} \leq \frac{n}{V_i(n)} \leq \frac{S_i^{(1)} + \ldots + S_i^{(V_i(n))}}{V_i(n)}. \tag{1.8}$$

By the strong law of large numbers

$$\mathbb{P}\left(\frac{S_i^{(1)} + \ldots + S_i^{(n)}}{n} \to m_i \text{ as } n \to \infty\right) = 1$$

and, since P is recurrent

$$\mathbb{P}(V_i(n) \to \infty \text{ as } n \to \infty) = 1.$$

So, letting $n \to \infty$ in (1.8), we get

$$\mathbb{P}\left(\frac{n}{V_i(n)} \to m_i \text{ as } n \to \infty\right) = 1,$$

which implies

$$\mathbb{P}\left(\frac{V_i(n)}{n} \to \frac{1}{m_i} \text{ as } n \to \infty\right) = 1.$$

Assume now that $(X_n)_{n\geq 0}$ has an invariant distribution $(\pi_i : i \in I)$. Let $f : I \to \mathbb{R}$ be a bounded function and assume without loss of generality that $|f| \leq 1$. For any $J \subseteq I$ we have

$$\left|\frac{1}{n}\sum_{k=0}^{n-1} f(X_k) - \overline{f}\right| = \left|\sum_{i\in I}\left(\frac{V_i(n)}{n} - \pi_i\right)f_i\right|$$

$$\leq \sum_{i\in J}\left|\frac{V_i(n)}{n} - \pi_i\right| + \sum_{i\notin J}\left|\frac{V_i(n)}{n} - \pi_i\right|$$

$$\leq \sum_{i\in J}\left|\frac{V_i(n)}{n} - \pi_i\right| + \sum_{i\notin J}\left(\frac{V_i(n)}{n} + \pi_i\right)$$

$$\leq 2\sum_{i\in J}\left|\frac{V_i(n)}{n} - \pi_i\right| + 2\sum_{i\notin J}\pi_i.$$

We proved above that

$$\mathbb{P}\left(\frac{V_i(n)}{n} \to \pi_i \text{ as } n \to \infty \text{ for all } i\right) = 1.$$

Given $\varepsilon > 0$, choose J finite so that

$$\sum_{i\notin J}\pi_i < \varepsilon/4$$

and then $N = N(\omega)$ so that, for $n \geq N(\omega)$

$$\sum_{i\in J}\left|\frac{V_i(n)}{n} - \pi_i\right| < \varepsilon/4.$$

Then, for $n \geq N(\omega)$, we have

$$\left|\frac{1}{n}\sum_{k=0}^{n-1} f(X_k) - \overline{f}\right| < \varepsilon,$$

which establishes the desired convergence. □

We consider now the statistical problem of estimating an unknown transition matrix P on the basis of observations of the corresponding Markov chain. Consider, to begin, the case where we have $N + 1$ observations $(X_n)_{0\leq n\leq N}$. The log-likelihood function is given by

$$l(P) = \log(\lambda_{X_0} p_{X_0 X_1} \ldots p_{X_{N-1} X_N}) = \sum_{i,j\in I} N_{ij}\log p_{ij}$$

up to a constant independent of P, where N_{ij} is the number of transitions from i to j. A standard statistical procedure is to find the *maximum likelihood estimate* $\widehat{P}$, which is the choice of P maximizing $l(P)$. Since P must satisfy the linear constraint $\sum_j p_{ij} = 1$ for each i, we first try to maximize

$$l(P) + \sum_{i,j \in I} \mu_i p_{ij}$$

and then choose $(\mu_i : i \in I)$ to fit the constraints. This is the method of Lagrange multipliers. Thus we find

$$\widehat{p}_{ij} = \sum_{n=0}^{N-1} 1_{\{X_n = i, X_{n+1} = j\}} / \sum_{n=0}^{N-1} 1_{\{X_n = i\}}$$

which is the proportion of jumps from i which go to j.

We now turn to consider the *consistency* of this sort of estimate, that is to say whether $\widehat{p}_{ij} \to p_{ij}$ with probability 1 as $N \to \infty$. Since this is clearly false when i is transient, we shall slightly modify our approach. Note that to find $\widehat{p}_{ij}$ we simply have to maximize

$$\sum_{j \in I} N_{ij} \log p_{ij}$$

subject to $\sum_j p_{ij} = 1$: the other terms and constraints are irrelevant. Suppose then that instead of $N + 1$ observations we make enough observations to ensure the chain leaves state i a total of N times. In the transient case this may involve restarting the chain several times. Denote again by N_{ij} the number of transitions from i to j.

To maximize the likelihood for $(p_{ij} : j \in I)$ we still maximize

$$\sum_{j \in I} N_{ij} \log p_{ij}$$

subject to $\sum_j p_{ij} = 1$, which leads to the maximum likelihood estimate

$$\widehat{p}_{ij} = N_{ij}/N.$$

But $N_{ij} = Y_1 + \ldots + Y_N$, where $Y_n = 1$ if the nth transition from i is to j, and $Y_n = 0$ otherwise. By the strong Markov property $Y_1, \ldots, Y_N$ are independent and identically distributed random variables with mean p_{ij}. So, by the strong law of large numbers

$$\mathbb{P}(\widehat{p}_{ij} \to p_{ij} \text{ as } N \to \infty) = 1,$$

which shows that $\widehat{p}_{ij}$ is consistent.

Exercises

1.10.1 Prove the claim (d) made in example (v) of the Introduction.

1.10.2 A professor has N umbrellas. He walks to the office in the morning and walks home in the evening. If it is raining he likes to carry an umbrella and if it is fine he does not. Suppose that it rains on each journey with probability p, independently of past weather. What is the long-run proportion of journeys on which the professor gets wet?

1.10.3 Let $(X_n)_{n\geq 0}$ be an irreducible Markov chain on I having an invariant distribution π. For $J \subseteq I$ let $(Y_m)_{m\geq 0}$ be the Markov chain on J obtained by observing $(X_n)_{n\geq 0}$ whilst in J. (See Example 1.4.4.) Show that $(Y_m)_{m\geq 0}$ is positive recurrent and find its invariant distribution.

1.10.4 An opera singer is due to perform a long series of concerts. Having a fine artistic temperament, she is liable to pull out each night with probability $1/2$. Once this has happened she will not sing again until the promoter convinces her of his high regard. This he does by sending flowers every day until she returns. Flowers costing x thousand pounds, $0 \leq x \leq 1$, bring about a reconciliation with probability $\sqrt{x}$. The promoter stands to make £750 from each successful concert. How much should he spend on flowers?

1.11 Appendix: recurrence relations

Recurrence relations often arise in the linear equations associated to Markov chains. Here is an account of the simplest cases. A more specialized case was dealt with in Example 1.3.4. In Example 1.1.4 we found a recurrence relation of the form
$$x_{n+1} = ax_n + b.$$
We look first for a constant solution $x_n = x$; then $x = ax + b$, so provided $a \neq 1$ we must have $x = b/(1-a)$. Now $y_n = x_n - b/(1-a)$ satisfies $y_{n+1} = ay_n$, so $y_n = a^n y_0$. Thus the general solution when $a \neq 1$ is given by
$$x_n = Aa^n + b/(1-a)$$
where A is a constant. When $a = 1$ the general solution is obviously
$$x_n = x_0 + nb.$$

In Example 1.3.3 we found a recurrence relation of the form
$$ax_{n+1} + bx_n + cx_{n-1} = 0$$

where a and c were both non-zero. Let us try a solution of the form $x_n = \lambda^n$; then $a\lambda^2 + b\lambda + c = 0$. Denote by α and β the roots of this quadratic. Then

$$y_n = A\alpha^n + B\beta^n$$

is a solution. If $\alpha \neq \beta$ then we can solve the equations

$$x_0 = A + B, \quad x_1 = A\alpha + B\beta$$

so that $y_0 = x_0$ and $y_1 = x_1$; but

$$a(y_{n+1} - x_{n+1}) + b(y_n - x_n) + c(y_{n-1} - x_{n-1}) = 0$$

for all n, so by induction $y_n = x_n$ for all n. If $\alpha = \beta \neq 0$, then

$$y_n = (A + nB)\alpha^n$$

is a solution and we can solve

$$x_0 = A\alpha^n, \quad x_1 = (A + B)\alpha^n$$

so that $y_0 = x_0$ and $y_1 = x_1$; then, by the same argument, $y_n = x_n$ for all n. The case $\alpha = \beta = 0$ does not arise. Hence the general solution is given by

$$x_n = \begin{cases} A\alpha^n + B\beta^n & \text{if } \alpha \neq \beta \\ (A + nB)\alpha^n & \text{if } \alpha = \beta. \end{cases}$$

1.12 Appendix: asymptotics for $n!$

Our analysis of recurrence and transience for random walks in Section 1.6 rested heavily on the use of the asymptotic relation

$$n! \sim A\sqrt{n}(n/e)^n \quad \text{as } n \to \infty$$

for some $A \in [1, \infty)$. Here is a derivation.

We make use of the power series expansions for $|t| < 1$

$$\log(1 + t) = t - \tfrac{1}{2}t^2 + \tfrac{1}{3}t^3 - \cdots$$
$$\log(1 - t) = -t - \tfrac{1}{2}t^2 - \tfrac{1}{3}t^3 - \cdots.$$

By subtraction we obtain

$$\tfrac{1}{2}\log\left(\frac{1+t}{1-t}\right) = t + \tfrac{1}{3}t^3 + \tfrac{1}{5}t^5 + \cdots.$$

1.12 Appendix: asymptotics for n!

Set $A_n = n!/(n^{n+1/2}e^{-n})$ and $a_n = \log A_n$. Then, by a straightforward calculation

$$a_n - a_{n+1} = (2n+1)\frac{1}{2}\log\left(\frac{1+(2n+1)^{-1}}{1-(2n+1)^{-1}}\right) - 1.$$

By the series expansion written above we have

$$\begin{aligned}
a_n - a_{n+1} &= (2n+1)\left\{\frac{1}{(2n+1)} + \frac{1}{3}\frac{1}{(2n+1)^3} + \frac{1}{5}\frac{1}{(2n+1)^5} + \cdots\right\} - 1 \\
&= \frac{1}{3}\frac{1}{(2n+1)^2} + \frac{1}{5}\frac{1}{(2n+1)^4} + \cdots \\
&\leq \frac{1}{3}\left\{\frac{1}{(2n+1)^2} + \frac{1}{(2n+1)^4} + \cdots\right\} \\
&= \frac{1}{3}\frac{1}{(2n+1)^2-1} = \frac{1}{12n} - \frac{1}{12(n+1)}.
\end{aligned}$$

It follows that a_n decreases and $a_n - 1/(12n)$ increases as $n \to \infty$. Hence $a_n \to a$ for some $a \in [0, \infty)$ and hence $A_n \to A$, as $n \to \infty$, where $A = e^a$.

2

Continuous-time Markov chains I

The material on continuous-time Markov chains is divided between this chapter and the next. The theory takes some time to set up, but once up and running it follows a very similar pattern to the discrete-time case. To emphasise this we have put the setting-up in this chapter and the rest in the next. If you wish, you can begin with Chapter 3, provided you take certain basic properties on trust, which are reviewed in Section 3.1. The first three sections of Chapter 2 fill in some necessary background information and are independent of each other. Section 2.4 on the Poisson process and Section 2.5 on birth processes provide a gentle warm-up for general continuous-time Markov chains. These processes are simple and particularly important examples of continuous-time chains. Sections 2.6–2.8, especially 2.8, deal with the heart of the continuous-time theory. There is an irreducible level of difficulty at this point, so we advise that Sections 2.7 and 2.8 are read selectively at first. Some examples of more general processes are given in Section 2.9. As in Chapter 1 the exercises form an important part of the text.

2.1 Q-matrices and their exponentials

In this section we shall discuss some of the basic properties of Q-matrices and explain their connection with continuous-time Markov chains.

Let I be a countable set. A *Q-matrix* on I is a matrix $Q = (q_{ij} : i, j \in I)$ satisfying the following conditions:

2.1 Q-matrices and their exponentials

(i) $0 \le -q_{ii} < \infty$ for all i;
(ii) $q_{ij} \ge 0$ for all $i \ne j$;
(iii) $\sum_{j \in I} q_{ij} = 0$ for all i.

Thus in each row of Q we can choose the off-diagonal entries to be any non-negative real numbers, subject only to the constraint that the off-diagonal row sum is finite:
$$q_i = \sum_{j \ne i} q_{ij} < \infty.$$

The diagonal entry q_{ii} is then $-q_i$, making the total row sum zero.

A convenient way to present the data for a continuous-time Markov chain is by means of a diagram, for example:

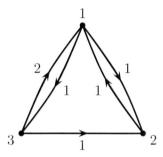

Each diagram then corresponds to a unique Q-matrix, in this case
$$Q = \begin{pmatrix} -2 & 1 & 1 \\ 1 & -1 & 0 \\ 2 & 1 & -3 \end{pmatrix}.$$

Thus each off-diagonal entry q_{ij} gives the value we attach to the (i, j) arrow on the diagram, which we shall interpret later as the *rate of going from i to j*. The numbers q_i are not shown on the diagram, but you can work them out from the other information given. We shall later interpret q_i as the *rate of leaving i*.

We may think of the discrete parameter space $\{0, 1, 2, \dots\}$ as embedded in the continuous parameter space $[0, \infty)$. For $p \in (0, \infty)$ a natural way to interpolate the discrete sequence $(p^n : n = 0, 1, 2, \dots)$ is by the function $(e^{tq} : t \ge 0)$, where $q = \log p$. Consider now a *finite* set I and a matrix

$P = (p_{ij} : i, j \in I)$. Is there a natural way to fill in the gaps in the discrete sequence $(P^n : n = 0, 1, 2, \dots)$?

For any matrix $Q = (q_{ij} : i, j \in I)$, the series

$$\sum_{k=0}^{\infty} \frac{Q^k}{k!}$$

converges componentwise and we denote its limit by e^Q. Moreover, if two matrices Q_1 and Q_2 commute, then

$$e^{Q_1+Q_2} = e^{Q_1} e^{Q_2}.$$

The proofs of these assertions follow the scalar case closely and are given in Section 2.10. Suppose then that we can find a matrix Q with $e^Q = P$. Then

$$e^{nQ} = (e^Q)^n = P^n$$

so $(e^{tQ} : t \geq 0)$ fills in the gaps in the discrete sequence.

Theorem 2.1.1. *Let Q be a matrix on a finite set I. Set $P(t) = e^{tQ}$. Then $(P(t) : t \geq 0)$ has the following properties:*

(i) $P(s+t) = P(s)P(t)$ *for all s, t (semigroup property);*

(ii) $(P(t) : t \geq 0)$ *is the unique solution to the forward equation*

$$\frac{d}{dt} P(t) = P(t)Q, \qquad P(0) = I ;$$

(iii) $(P(t) : t \geq 0)$ *is the unique solution to the backward equation*

$$\frac{d}{dt} P(t) = QP(t), \qquad P(0) = I ;$$

(iv) *for $k = 0, 1, 2, \dots$, we have*

$$\left(\frac{d}{dt}\right)^k \bigg|_{t=0} P(t) = Q^k.$$

Proof. For any $s, t \in \mathbb{R}$, sQ and tQ commute, so

$$e^{sQ} e^{tQ} = e^{(s+t)Q}$$

proving the semigroup property. The matrix-valued power series

$$P(t) = \sum_{k=0}^{\infty} \frac{(tQ)^k}{k!}$$

has infinite radius of convergence (see Section 2.10). So each component is differentiable with derivative given by term-by-term differentiation:

$$P'(t) = \sum_{k=1}^{\infty} \frac{t^{k-1}Q^k}{(k-1)!} = P(t)Q = QP(t).$$

Hence $P(t)$ satisfies the forward and backward equations. Moreover by repeated term-by-term differentiation we obtain (iv). It remains to show that $P(t)$ is the only solution of the forward and backward equations. But if $M(t)$ satisfies the forward equation, then

$$\frac{d}{dt}(M(t)e^{-tQ}) = \left(\frac{d}{dt}M(t)\right)e^{-tQ} + M(t)\left(\frac{d}{dt}e^{-tQ}\right)$$
$$= M(t)Qe^{-tQ} + M(t)(-Q)e^{-tQ} = 0$$

so $M(t)e^{-tQ}$ is constant, and so $M(t) = P(t)$. A similar argument proves uniqueness for the backward equation. □

The last result was about matrix exponentials in general. Now let us see what happens to Q-matrices. Recall that a matrix $P = (p_{ij} : i, j \in I)$ is stochastic if it satisfies

(i) $0 \leq p_{ij} < \infty$ for all i, j;
(ii) $\sum_{j \in I} p_{ij} = 1$ for all i.

We recall the convention that in the limit $t \to 0$ the statement $f(t) = O(t)$ means that $f(t)/t \leq C$ for all sufficiently small t, for some $C < \infty$. Later we shall also use the convention that $f(t) = o(t)$ means $f(t)/t \to 0$ as $t \to 0$.

Theorem 2.1.2. *A matrix Q on a finite set I is a Q-matrix if and only if $P(t) = e^{tQ}$ is a stochastic matrix for all $t \geq 0$.*

Proof. As $t \downarrow 0$ we have

$$P(t) = I + tQ + O(t^2)$$

so $q_{ij} \geq 0$ for $i \neq j$ if and only if $p_{ij}(t) \geq 0$ for all i, j and $t \geq 0$ sufficiently small. Since $P(t) = P(t/n)^n$ for all n, it follows that $q_{ij} \geq 0$ for $i \neq j$ if and only if $p_{ij}(t) \geq 0$ for all i, j and *all* $t \geq 0$.

If Q has zero row sums then so does Q^n for every n:

$$\sum_{k \in I} q_{ik}^{(n)} = \sum_{k \in I} \sum_{j \in I} q_{ij}^{(n-1)} q_{jk} = \sum_{j \in I} q_{ij}^{(n-1)} \sum_{k \in I} q_{jk} = 0.$$

So
$$\sum_{j \in I} p_{ij}(t) = 1 + \sum_{n=1}^{\infty} \frac{t^n}{n!} \sum_{j \in I} q_{ij}^{(n)} = 1.$$

On the other hand, if $\sum_{j \in I} p_{ij}(t) = 1$ for all $t \geq 0$, then
$$\sum_{j \in I} q_{ij} = \frac{d}{dt}\bigg|_{t=0} \sum_{j \in I} p_{ij}(t) = 0. \qquad \square$$

Now, if P is a stochastic matrix of the form e^Q for some Q-matrix, we can do some sort of filling-in of gaps at the level of processes. Fix some large integer m and let $(X_n^m)_{n \geq 0}$ be discrete-time Markov$(\lambda, e^{Q/m})$. We define a process indexed by $\{n/m : n = 0, 1, 2, \dots\}$ by
$$X_{n/m} = X_n^m.$$

Then $(X_n : n = 0, 1, 2, \dots)$ is discrete-time Markov$(\lambda, (e^{Q/m})^m)$ (see Exercise 1.1.2) and
$$(e^{Q/m})^m = e^Q = P.$$

Thus we can find discrete-time Markov chains with arbitrarily fine grids $\{n/m : n = 0, 1, 2, \dots\}$ as time-parameter sets which give rise to Markov(λ, P) when sampled at integer times. It should not then be too surprising that there is, as we shall see in Section 2.8, a continuous-time process $(X_t)_{t \geq 0}$ which also has this property.

To anticipate a little, we shall see in Section 2.8 that a continuous-time Markov chain $(X_t)_{t \geq 0}$ with Q-matrix Q satisfies
$$\mathbb{P}(X_{t_{n+1}} = i_{n+1} \mid X_{t_0} = i_0, \dots, X_{t_n} = i_n) = p_{i_n i_{n+1}}(t_{n+1} - t_n)$$

for all $n = 0, 1, 2, \dots$, all times $0 \leq t_0 \leq \dots \leq t_{n+1}$ and all states $i_0, \dots, i_{n+1}$, where $p_{ij}(t)$ is the (i, j) entry in e^{tQ}. In particular, the *transition probability* from i to j in time t is given by
$$\mathbb{P}_i(X_t = j) := \mathbb{P}(X_t = j \mid X_0 = i) = p_{ij}(t).$$

(Recall that := means 'defined to equal'.) You should compare this with the defining property of a discrete-time Markov chain given in Section 1.1. We shall now give some examples where the transition probabilities $p_{ij}(t)$ may be calculated explicitly.

Example 2.1.3

We calculate $p_{11}(t)$ for the continuous-time Markov chain with Q-matrix

$$Q = \begin{pmatrix} -2 & 1 & 1 \\ 1 & -1 & 0 \\ 2 & 1 & -3 \end{pmatrix}.$$

The method is similar to that of Example 1.1.6. We begin by writing down the characteristic equation for Q:

$$0 = \det(x - Q) = x(x+2)(x+4).$$

This shows that Q has distinct eigenvalues $0, -2, -4$. Then $p_{11}(t)$ has the form

$$p_{11}(t) = a + be^{-2t} + ce^{-4t}$$

for some constants a, b and c. (This is because we could diagonalize Q by an invertible matrix U:

$$Q = U \begin{pmatrix} 0 & 0 & 0 \\ 0 & -2 & 0 \\ 0 & 0 & -4 \end{pmatrix} U^{-1}.$$

Then

$$e^{tQ} = \sum_{k=0}^{\infty} \frac{(tQ)^k}{k!}$$

$$= U \sum_{k=0}^{\infty} \frac{1}{k!} \begin{pmatrix} 0^k & 0 & 0 \\ 0 & (-2t)^k & 0 \\ 0 & 0 & (-4t)^k \end{pmatrix} U^{-1}$$

$$= U \begin{pmatrix} 1 & 0 & 0 \\ 0 & e^{-2t} & 0 \\ 0 & 0 & e^{-4t} \end{pmatrix} U^{-1},$$

so $p_{11}(t)$ must have the form claimed.) To determine the constants we use

$$1 = p_{11}(0) = a + b + c,$$
$$-2 = q_{11} = p'_{11}(0) = -2b - 4c,$$
$$7 = q_{11}^{(2)} = p''_{11}(0) = 4b + 16c,$$

so

$$p_{11}(t) = \tfrac{3}{8} + \tfrac{1}{4}e^{-2t} + \tfrac{3}{8}e^{-4t}.$$

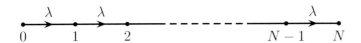

Example 2.1.4

We calculate $p_{ij}(t)$ for the continuous-time Markov chain with diagram given above. The Q-matrix is

$$Q = \begin{pmatrix} -\lambda & \lambda & & & & \\ & -\lambda & \lambda & & & \\ & & \ddots & \ddots & & \\ & & & & \lambda & \\ & & & & -\lambda & \lambda \\ & & & & & 0 \end{pmatrix}$$

where entries off the diagonal and super-diagonal are all zero. The exponential of an upper-triangular matrix is upper-triangular, so $p_{ij}(t) = 0$ for $i > j$. In components the forward equation $P'(t) = P(t)Q$ reads

$$p'_{ii}(t) = -\lambda p_{ii}(t), \qquad p_{ii}(0) = 1, \qquad \text{for } i < N,$$
$$p'_{ij}(t) = -\lambda p_{ij}(t) + \lambda p_{i,j-1}(t), \qquad p_{ij}(0) = 0, \qquad \text{for } i < j < N,$$
$$p'_{iN}(t) = \lambda p_{iN-1}(t), \qquad p_{iN}(0) = 0, \qquad \text{for } i < N.$$

We can solve these equations. First, $p_{ii}(t) = e^{-\lambda t}$ for $i < N$. Then, for $i < j < N$

$$(e^{\lambda t} p_{ij}(t))' = e^{\lambda t} p_{i,j-1}(t)$$

so, by induction

$$p_{ij}(t) = e^{-\lambda t} \frac{(\lambda t)^{j-i}}{(j-i)!}.$$

If $i = 0$, these are the Poisson probabilities of parameter λt. So, starting from 0, the distribution of the Markov chain at time t is the same as the distribution of $\min\{Y_t, N\}$, where Y_t is a Poisson random variable of parameter λt.

Exercises

2.1.1 Compute $p_{11}(t)$ for $P(t) = e^{tQ}$, where

$$Q = \begin{pmatrix} -2 & 1 & 1 \\ 4 & -4 & 0 \\ 2 & 1 & -3 \end{pmatrix}.$$

2.1.2 Which of the following matrices is the exponential of a Q-matrix?

$$\text{(a)} \begin{pmatrix} 1 & 0 \\ 0 & 1 \end{pmatrix} \quad \text{(b)} \begin{pmatrix} 1 & 0 \\ 1 & 0 \end{pmatrix} \quad \text{(c)} \begin{pmatrix} 0 & 1 \\ 1 & 0 \end{pmatrix}.$$

What consequences do your answers have for the discrete-time Markov chains with these transition matrices?

2.2 Continuous-time random processes

Let I be a countable set. A *continuous-time random process*

$$(X_t)_{t\geq 0} = (X_t : 0 \leq t < \infty)$$

with values in I is a family of random variables $X_t : \Omega \to I$. We are going to consider ways in which we might specify the probabilistic behaviour (or *law*) of $(X_t)_{t\geq 0}$. These should enable us to find, at least in principle, any probability connected with the process, such as $\mathbb{P}(X_t = i)$ or $\mathbb{P}(X_{t_0} = i_0, \ldots, X_{t_n} = i_n)$, or $\mathbb{P}(X_t = i \text{ for some } t)$. There are subtleties in this problem not present in the discrete-time case. They arise because, for a countable disjoint union

$$\mathbb{P}\left(\bigcup_n A_n\right) = \sum_n \mathbb{P}(A_n),$$

whereas for an uncountable union $\bigcup_{t\geq 0} A_t$ there is no such rule. To avoid these subtleties as far as possible we shall restrict our attention to processes $(X_t)_{t\geq 0}$ which are *right-continuous*. This means in this context that for all $\omega \in \Omega$ and $t \geq 0$ there exists $\varepsilon > 0$ such that

$$X_s(\omega) = X_t(\omega) \quad \text{for } t \leq s \leq t + \varepsilon.$$

By a standard result of measure theory, which is proved in Section 6.6, the probability of any event depending on a right-continuous process can be determined from its *finite-dimensional distributions*, that is, from the probabilities

$$\mathbb{P}(X_{t_0} = i_0, X_{t_1} = i_1, \ldots, X_{t_n} = i_n)$$

for $n \geq 0$, $0 \leq t_0 \leq t_1 \leq \ldots \leq t_n$ and $i_0, \ldots, i_n \in I$. For example

$$\mathbb{P}(X_t = i \text{ for some } t \in [0, \infty)) = 1 - \lim_{n\to\infty} \sum_{j_1,\ldots,j_n \neq i} \mathbb{P}(X_{q_1} = j_i, \ldots, X_{q_n} = j_n)$$

where $q_1, q_2, \ldots$ is an enumeration of the rationals.

68   2. Continuous-time Markov chains I

Every path $t \mapsto X_t(\omega)$ of a right-continuous process must remain constant for a while in each new state, so there are three possibilities for the sorts of path we get. In the first case the path makes infinitely many jumps, but only finitely many in any interval $[0, t]$:

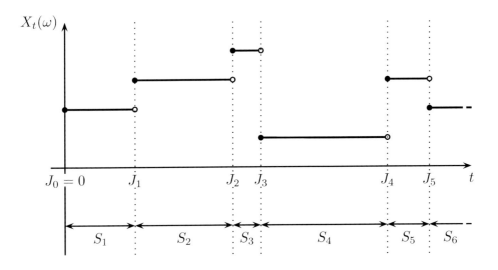

The second case is where the path makes finitely many jumps and then becomes stuck in some state forever:

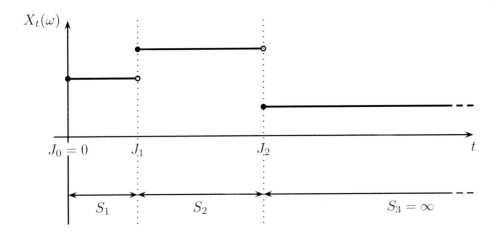

In the third case the process makes infinitely many jumps in a finite interval; this is illustrated below. In this case, after the explosion time ζ the process starts up again; it may explode again, maybe infinitely often, or it may not.

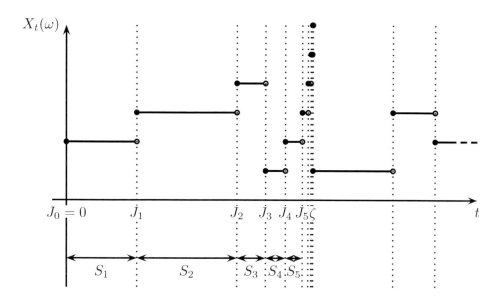

We call $J_0, J_1, \ldots$ the *jump times* of $(X_t)_{t \geq 0}$ and $S_1, S_2, \ldots$ the *holding times*. They are obtained from $(X_t)_{t \geq 0}$ by

$$J_0 = 0, \quad J_{n+1} = \inf\{t \geq J_n : X_t \neq X_{J_n}\}$$

for $n = 0, 1, \ldots$, where $\inf \emptyset = \infty$, and, for $n = 1, 2, \ldots$,

$$S_n = \begin{cases} J_n - J_{n-1} & \text{if } J_{n-1} < \infty \\ \infty & \text{otherwise.} \end{cases}$$

Note that right-continuity forces $S_n > 0$ for all n. If $J_{n+1} = \infty$ for some n, we define $X_\infty = X_{J_n}$, the final value, otherwise X_∞ is undefined. The *(first) explosion time* ζ is defined by

$$\zeta = \sup_n J_n = \sum_{n=1}^{\infty} S_n.$$

The discrete-time process $(Y_n)_{n \geq 0}$ given by $Y_n = X_{J_n}$ is called the *jump process* of $(X_t)_{t \geq 0}$, or the *jump chain* if it is a discrete-time Markov chain. This is simply the sequence of values taken by $(X_t)_{t \geq 0}$ up to explosion.

We shall not consider what happens to a process after explosion. So it is convenient to adjoin to I a new state, ∞ say, and require that $X_t = \infty$ if $t \geq \zeta$. Any process satisfying this requirement is called *minimal*. The terminology 'minimal' does not refer to the state of the process but to the interval of time over which the process is active. Note that a minimal process may be reconstructed from its holding times and jump process. Thus by specifying the joint distribution of $S_1, S_2, \ldots$ and $(Y_n)_{n \geq 0}$ we have another 'countable' specification of the probabilistic behaviour of $(X_t)_{t \geq 0}$. For example, the probability that $X_t = i$ is given by

$$\mathbb{P}(X_t = i) = \sum_{n=0}^{\infty} \mathbb{P}(Y_n = i \text{ and } J_n \leq t < J_{n+1})$$

and

$$\mathbb{P}(X_t = i \text{ for some } t \in [0, \infty)) = \mathbb{P}(Y_n = i \text{ for some } n \geq 0).$$

2.3 Some properties of the exponential distribution

A random variable $T : \Omega \to [0, \infty]$ has *exponential distribution of parameter* λ $(0 \leq \lambda < \infty)$ if
$$\mathbb{P}(T > t) = e^{-\lambda t} \quad \text{for all } t \geq 0.$$

We write $T \sim E(\lambda)$ for short. If $\lambda > 0$, then T has density function

$$f_T(t) = \lambda e^{-\lambda t} 1_{t \geq 0}.$$

The mean of T is given by

$$\mathbb{E}(T) = \int_0^{\infty} \mathbb{P}(T > t) dt = \lambda^{-1}.$$

The exponential distribution plays a fundamental role in continuous-time Markov chains because of the following results.

Theorem 2.3.1 (Memoryless property). *A random variable $T : \Omega \to (0, \infty]$ has an exponential distribution if and only if it has the following memoryless property:*

$$\mathbb{P}(T > s + t \mid T > s) = \mathbb{P}(T > t) \quad \text{for all } s, t \geq 0.$$

Proof. Suppose $T \sim E(\lambda)$, then

$$\mathbb{P}(T > s + t \mid T > s) = \frac{\mathbb{P}(T > s + t)}{\mathbb{P}(T > s)} = \frac{e^{-\lambda(s+t)}}{e^{-\lambda s}} = e^{-\lambda t} = \mathbb{P}(T > t).$$

On the other hand, suppose T has the memoryless property whenever $\mathbb{P}(T > s) > 0$. Then $g(t) = \mathbb{P}(T > t)$ satisfies
$$g(s+t) = g(s)g(t) \quad \text{for all } s, t \geq 0.$$

We assumed $T > 0$ so that $g(1/n) > 0$ for some n. Then, by induction
$$g(1) = g\left(\frac{1}{n} + \ldots + \frac{1}{n}\right) = g\left(\frac{1}{n}\right)^n > 0$$

so $g(1) = e^{-\lambda}$ for some $0 \leq \lambda < \infty$. By the same argument, for integers $p, q \geq 1$
$$g(p/q) = g(1/q)^p = g(1)^{p/q}$$

so $g(r) = e^{-\lambda r}$ for all rationals $r > 0$. For real $t > 0$, choose rationals $r, s > 0$ with $r \leq t \leq s$. Since g is decreasing,
$$e^{-\lambda r} = g(r) \geq g(t) \geq g(s) = e^{-\lambda s}$$

and, since we can choose r and s arbitrarily close to t, this forces $g(t) = e^{-\lambda t}$, so $T \sim E(\lambda)$. $\square$

The next result shows that a sum of independent exponential random variables is either certain to be finite or certain to be infinite, and gives a criterion for deciding which is true. This will be used to determine whether or not certain continuous-time Markov chains can take infinitely many jumps in a finite time.

Theorem 2.3.2. *Let $S_1, S_2, \ldots$ be a sequence of independent random variables with $S_n \sim E(\lambda_n)$ and $0 < \lambda_n < \infty$ for all n.*

(i) *If $\displaystyle\sum_{n=1}^{\infty} \frac{1}{\lambda_n} < \infty$, then $\displaystyle\mathbb{P}\left(\sum_{n=1}^{\infty} S_n < \infty\right) = 1$.*

(ii) *If $\displaystyle\sum_{n=1}^{\infty} \frac{1}{\lambda_n} = \infty$, then $\displaystyle\mathbb{P}\left(\sum_{n=1}^{\infty} S_n = \infty\right) = 1$.*

Proof. (i) Suppose $\sum_{n=1}^{\infty} 1/\lambda_n < \infty$. Then, by monotone convergence
$$\mathbb{E}\left(\sum_{n=1}^{\infty} S_n\right) = \sum_{n=1}^{\infty} \frac{1}{\lambda_n} < \infty$$

so
$$\mathbb{P}\left(\sum_{n=1}^{\infty} S_n < \infty\right) = 1.$$

(ii) Suppose instead that $\sum_{n=1}^{\infty} 1/\lambda_n = \infty$. Then $\prod_{n=1}^{\infty}(1 + 1/\lambda_n) = \infty$. By monotone convergence and independence

$$\mathbb{E}\left(\exp\left\{-\sum_{n=1}^{\infty} S_n\right\}\right) = \prod_{n=1}^{\infty} \mathbb{E}\left(\exp\{-S_n\}\right) = \prod_{n=1}^{\infty}\left(1 + \frac{1}{\lambda_n}\right)^{-1} = 0$$

so

$$\mathbb{P}\left(\sum_{n=1}^{\infty} S_n = \infty\right) = 1. \qquad \square$$

The following result is fundamental to continuous-time Markov chains.

Theorem 2.3.3. *Let I be a countable set and let $T_k, k \in I$, be independent random variables with $T_k \sim E(q_k)$ and $0 < q := \sum_{k \in I} q_k < \infty$. Set $T = \inf_k T_k$. Then this infimum is attained at a unique random value K of k, with probability 1. Moreover, T and K are independent, with $T \sim E(q)$ and $\mathbb{P}(K = k) = q_k/q$.*

Proof. Set $K = k$ if $T_k < T_j$ for all $j \neq k$, otherwise let K be undefined. Then

$$\mathbb{P}(K = k \text{ and } T \geq t)$$
$$= \mathbb{P}(T_k \geq t \text{ and } T_j > T_k \text{ for all } j \neq k)$$
$$= \int_t^{\infty} q_k e^{-q_k s} \mathbb{P}(T_j > s \text{ for all } j \neq k) ds$$
$$= \int_t^{\infty} q_k e^{-q_k s} \prod_{j \neq k} e^{-q_j s} ds$$
$$= \int_t^{\infty} q_k e^{-qs} ds = \frac{q_k}{q} e^{-qt}.$$

Hence $\mathbb{P}(K = k \text{ for some } k) = 1$ and T and K have the claimed joint distribution. $\square$

The following identity is the simplest case of an identity used in Section 2.8 in proving the forward equations for a continuous-time Markov chain.

Theorem 2.3.4. *For independent random variables $S \sim E(\lambda)$ and $R \sim E(\mu)$ and for $t \geq 0$, we have*

$$\mu \mathbb{P}(S \leq t < S + R) = \lambda \mathbb{P}(R \leq t < R + S).$$

Proof. We have

$$\mu\mathbb{P}(S \leq t < S+R) = \mu \int_0^t \int_{t-s}^\infty \lambda\mu e^{-\lambda s} e^{-\mu r}\,dr\,ds = \lambda\mu \int_0^t e^{-\lambda s} e^{-\mu(t-s)}\,ds$$

from which the identity follows by symmetry. $\square$

Exercises

2.3.1 Suppose S and T are independent exponential random variables of parameters α and β respectively. What is the distribution of $\min\{S, T\}$? What is the probability that $S \leq T$? Show that the two events $\{S < T\}$ and $\{\min\{S, T\} \geq t\}$ are independent.

2.3.2 Let $T_1, T_2, \ldots$ be independent exponential random variables of parameter λ and let N be an independent geometric random variable with

$$\mathbb{P}(N = n) = \beta(1-\beta)^{n-1}, \qquad n = 1, 2, \ldots.$$

Show that $T = \sum_{i=1}^N T_i$ has exponential distribution of parameter $\lambda\beta$.

2.3.3 Let $S_1, S_2, \ldots$ be independent exponential random variables with parameters $\lambda_1, \lambda_2, \ldots$ respectively. Show that $\lambda_1 S_1$ is exponential of parameter 1.

Use the strong law of large numbers to show, first in the special case $\lambda_n = 1$ for all n, and then subject only to the condition $\sup_n \lambda_n < \infty$, that

$$\mathbb{P}\left(\sum_{n=1}^\infty S_n = \infty\right) = 1.$$

Is the condition $\sup_n \lambda_n < \infty$ absolutely necessary?

2.4 Poisson processes

Poisson processes are some of the simplest examples of continuous-time Markov chains. We shall also see that they may serve as building blocks for the most general continuous-time Markov chain. Moreover, a Poisson process is the natural probabilistic model for any uncoordinated stream of discrete events in continuous time. So we shall study Poisson processes first, both as a gentle warm-up for the general theory and because they are useful in themselves. The key result is Theorem 2.4.3, which provides three different descriptions of a Poisson process. The reader might well begin with the statement of this result and then see how it is used in the theorems and examples that follow. We shall begin with a definition in terms of jump chain and holding times (see Section 2.2). A right-continuous process $(X_t)_{t\geq 0}$ with values in $\{0, 1, 2, \dots\}$ is a *Poisson process of rate* λ $(0 < \lambda < \infty)$ if its holding times $S_1, S_2, \dots$ are independent exponential random variables of parameter λ and its jump chain is given by $Y_n = n$. Here is the diagram:

$$\begin{array}{ccccccccc} & \lambda & & \lambda & & \lambda & & \lambda & \\ \bullet & \to & \bullet & \to & \bullet & \to & \bullet & \to & \bullet \quad \text{---} \\ 0 & & 1 & & 2 & & 3 & & 4 \end{array}$$

The associated Q-matrix is given by

$$Q = \begin{pmatrix} -\lambda & \lambda & & & \\ & -\lambda & \lambda & & \\ & & \ddots & \ddots & \\ & & & & \end{pmatrix}.$$

By Theorem 2.3.2 (or the strong law of large numbers) we have $\mathbb{P}(J_n \to \infty) = 1$ so there is no explosion and the law of $(X_t)_{t\geq 0}$ is uniquely determined. A simple way to construct a Poisson process of rate λ is to take a sequence $S_1, S_2, \dots$ of independent exponential random variables of parameter λ, to set $J_0 = 0$, $J_n = S_1 + \dots + S_n$ and then set

$$X_t = n \quad \text{if} \quad J_n \leq t < J_{n+1}.$$

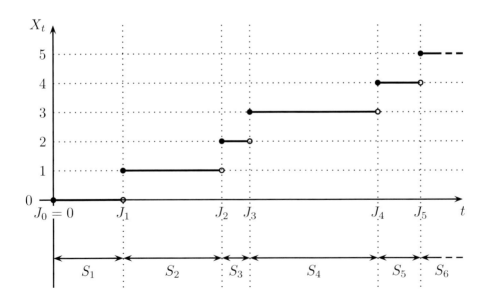

2.4 Poisson processes

The diagram illustrates a typical path. We now show how the memoryless property of the exponential holding times, Theorem 2.3.1, leads to a memoryless property of the Poisson process.

Theorem 2.4.1 (Markov property). Let $(X_t)_{t\geq 0}$ be a Poisson process of rate λ. Then, for any $s \geq 0$, $(X_{s+t} - X_s)_{t\geq 0}$ is also a Poisson process of rate λ, independent of $(X_r : r \leq s)$.

Proof. It suffices to prove the claim conditional on the event $X_s = i$, for each $i \geq 0$. Set $\widetilde{X}_t = X_{s+t} - X_s$. We have

$$\{X_s = i\} = \{J_i \leq s < J_{i+1}\} = \{J_i \leq s\} \cap \{S_{i+1} > s - J_i\}.$$

On this event

$$X_r = \sum_{j=1}^{i} 1_{\{S_j \leq r\}} \quad \text{for } r \leq s$$

and the holding times $\widetilde{S}_1, \widetilde{S}_2, \ldots$ of $(\widetilde{X}_t)_{t\geq 0}$ are given by

$$\widetilde{S}_1 = S_{i+1} - (s - J_i), \quad \widetilde{S}_n = S_{i+n} \quad \text{for } n \geq 2$$

as shown in the diagram.

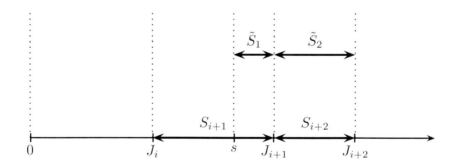

Recall that the holding times $S_1, S_2, \ldots$ are independent $E(\lambda)$. Condition on $S_1, \ldots, S_i$ and $\{X_s = i\}$, then by the memoryless property of S_{i+1} and independence, $\widetilde{S}_1, \widetilde{S}_2, \ldots$ are themselves independent $E(\lambda)$. Hence, conditional on $\{X_s = i\}$, $\widetilde{S}_1, \widetilde{S}_2, \ldots$ are independent $E(\lambda)$, and independent of $S_1, \ldots, S_i$. Hence, conditional on $\{X_s = i\}$, $(\widetilde{X}_t)_{t\geq 0}$ is a Poisson process of rate λ and independent of $(X_r : r \leq s)$. □

In fact, we shall see in Section 6.5, by an argument in essentially the same spirit that the result also holds with s replaced by any stopping time T of $(X_t)_{t\geq 0}$.

Theorem 2.4.2 (Strong Markov property). *Let $(X_t)_{t\geq 0}$ be a Poisson process of rate λ and let T be a stopping time of $(X_t)_{t\geq 0}$. Then, conditional on $T<\infty$, $(X_{T+t}-X_T)_{t\geq 0}$ is also a Poisson process of rate λ, independent of $(X_s : s \leq T)$.*

Here is some standard terminology. If $(X_t)_{t\geq 0}$ is a real-valued process, we can consider its *increment* $X_t - X_s$ over any interval $(s,t]$. We say that $(X_t)_{t\geq 0}$ has *stationary* increments if the distribution of $X_{s+t} - X_s$ depends only on $t \geq 0$. We say that $(X_t)_{t\geq 0}$ has *independent* increments if its increments over any finite collection of disjoint intervals are independent.

We come to the key result for the Poisson process, which gives two conditions equivalent to the jump chain/holding time characterization which we took as our original definition. Thus we have three alternative definitions of the same process.

Theorem 2.4.3. *Let $(X_t)_{t\geq 0}$ be an increasing, right-continuous integer-valued process starting from 0. Let $0 < \lambda < \infty$. Then the following three conditions are equivalent:*

(a) *(jump chain/holding time definition) the holding times $S_1, S_2, \ldots$ of $(X_t)_{t\geq 0}$ are independent exponential random variables of parameter λ and the jump chain is given by $Y_n = n$ for all n;*

(b) *(infinitesimal definition) $(X_t)_{t\geq 0}$ has independent increments and, as $h \downarrow 0$, uniformly in t,*

$$\mathbb{P}(X_{t+h} - X_t = 0) = 1 - \lambda h + o(h), \quad \mathbb{P}(X_{t+h} - X_t = 1) = \lambda h + o(h);$$

(c) *(transition probability definition) $(X_t)_{t\geq 0}$ has stationary independent increments and, for each t, X_t has Poisson distribution of parameter λt.*

If $(X_t)_{t\geq 0}$ satisfies any of these conditions then it is called a *Poisson process of rate λ*.

Proof. (a) $\Rightarrow$ (b) If (a) holds, then, by the Markov property, for any $t, h \geq 0$, the increment $X_{t+h} - X_t$ has the same distribution as X_h and is independent of $(X_s : s \leq t)$. So $(X_t)_{t\geq 0}$ has independent increments and as $h \downarrow 0$

$$\mathbb{P}(X_{t+h} - X_t \geq 1) = \mathbb{P}(X_h \geq 1) = \mathbb{P}(J_1 \leq h) = 1 - e^{-\lambda h} = \lambda h + o(h),$$
$$\mathbb{P}(X_{t+h} - X_t \geq 2) = \mathbb{P}(X_h \geq 2) = \mathbb{P}(J_2 \leq h)$$
$$\leq \mathbb{P}(S_1 \leq h \text{ and } S_2 \leq h) = (1 - e^{-\lambda h})^2 = o(h),$$

which implies (b).

(b) ⇒ (c) If (b) holds, then, for $i = 2, 3, \ldots$, we have $\mathbb{P}(X_{t+h} - X_t = i) = o(h)$ as $h \downarrow 0$, uniformly in t. Set $p_j(t) = \mathbb{P}(X_t = j)$. Then, for $j = 1, 2, \ldots$,

$$p_j(t+h) = \mathbb{P}(X_{t+h} = j) = \sum_{i=0}^{j} \mathbb{P}(X_{t+h} - X_t = i)\, \mathbb{P}(X_t = j - i)$$
$$= (1 - \lambda h + o(h)) p_j(t) + (\lambda h + o(h)) p_{j-1}(t) + o(h)$$

so

$$\frac{p_j(t+h) - p_j(t)}{h} = -\lambda p_j(t) + \lambda p_{j-1}(t) + O(h).$$

Since this estimate is uniform in t we can put $t = s - h$ to obtain for all $s \geq h$

$$\frac{p_j(s) - p_j(s-h)}{h} = -\lambda p_j(s-h) + \lambda p_{j-1}(s-h) + O(h).$$

Now let $h \downarrow 0$ to see that $p_j(t)$ is first continuous and then differentiable and satisfies the differential equation

$$p_j'(t) = -\lambda p_j(t) + \lambda p_{j-1}(t).$$

By a simpler argument we also find

$$p_0'(t) = -\lambda p_0(t).$$

Since $X_0 = 0$ we have initial conditions

$$p_0(0) = 1, \quad p_j(0) = 0 \quad \text{for } j = 1, 2, \ldots .$$

As we saw in Example 2.1.4, this system of equations has a unique solution given by

$$p_j(t) = e^{-\lambda t} \frac{(\lambda t)^j}{j!}, \quad j = 0, 1, 2, \ldots .$$

Hence $X_t \sim P(\lambda t)$. If $(X_t)_{t \geq 0}$ satisfies (b), then certainly $(X_t)_{t \geq 0}$ has independent increments, but also $(X_{s+t} - X_s)_{t \geq 0}$ satisfies (b), so the above argument shows $X_{s+t} - X_s \sim P(\lambda t)$, for any s, which implies (c).

(c) ⇒ (a) There is a process satisfying (a) and we have shown that it must then satisfy (c). But condition (c) determines the finite-dimensional distributions of $(X_t)_{t \geq 0}$ and hence the distribution of jump chain and holding times. So if one process satisfying (c) also satisfies (a), so must every process satisfying (c). □

The differential equations which appeared in the proof are really the forward equations for the Poisson process. To make this clear, consider the

possibility of starting the process from i at time 0, writing $\mathbb{P}_i$ as a reminder, and set
$$p_{ij}(t) = \mathbb{P}_i(X_t = j).$$
Then, by spatial homogeneity $p_{ij}(t) = p_{j-i}(t)$, and we could rewrite the differential equations as

$$p'_{i0}(t) = -\lambda p_{i0}(t), \qquad p_{i0}(0) = \delta_{i0},$$
$$p'_{ij}(t) = \lambda p_{i,j-1}(t) - \lambda p_{ij}(t), \qquad p_{ij}(0) = \delta_{ij}$$

or, in matrix form, for Q as above,
$$P'(t) = P(t)Q, \qquad P(0) = I.$$

Theorem 2.4.3 contains a great deal of information about the Poisson process of rate λ. It can be useful when trying to decide whether a given process is a Poisson process as it gives you three alternative conditions to check, and it is likely that one will be easier to check than another. On the other hand it can also be useful when answering a question about a given Poisson process as this question may be more closely connected to one definition than another. For example, you might like to consider the difficulties in approaching the next result using the jump chain/holding time definition.

Theorem 2.4.4. *If $(X_t)_{t\geq 0}$ and $(Y_t)_{t\geq 0}$ are independent Poisson processes of rates λ and μ, respectively, then $(X_t + Y_t)_{t\geq 0}$ is a Poisson process of rate $\lambda + \mu$.*

Proof. We shall use the infinitesimal definition, according to which $(X_t)_{t\geq 0}$ and $(Y_t)_{t\geq 0}$ have independent increments and, as $h \downarrow 0$, uniformly in t,

$$\mathbb{P}(X_{t+h} - X_t = 0) = 1 - \lambda h + o(h), \quad \mathbb{P}(X_{t+h} - X_t = 1) = \lambda h + o(h),$$
$$\mathbb{P}(Y_{t+h} - Y_t = 0) = 1 - \mu h + o(h), \quad \mathbb{P}(Y_{t+h} - Y_t = 1) = \mu h + o(h).$$

Set $Z_t = X_t + Y_t$. Then, since $(X_t)_{t\geq 0}$ and $(Y_t)_{t\geq 0}$ are independent, $(Z_t)_{t\geq 0}$ has independent increments and, as $h \downarrow 0$, uniformly in t,

$$\mathbb{P}(Z_{t+h} - Z_t = 0) = \mathbb{P}(X_{t+h} - X_t = 0)\mathbb{P}(Y_{t+h} - Y_t = 0)$$
$$= (1 - \lambda h + o(h))(1 - \mu h + o(h)) = 1 - (\lambda + \mu)h + o(h),$$
$$\mathbb{P}(Z_{t+h} - Z_t = 1) = \mathbb{P}(X_{t+h} - X_t = 1)\mathbb{P}(Y_{t+h} - Y_t = 0)$$
$$\qquad + \mathbb{P}(X_{t+h} - X_t = 0)\mathbb{P}(Y_{t+h} - Y_t = 1)$$
$$= (\lambda h + o(h))(1 - \mu h + o(h)) + (1 - \lambda h + o(h))(\mu h + o(h))$$
$$= (\lambda + \mu)h + o(h).$$

Hence $(Z_t)_{t\geq 0}$ is a Poisson process of rate $\lambda + \mu$. $\square$

Next we establish some relations between Poisson processes and the uniform distribution. Notice that the conclusions are independent of the rate of the process considered. The results say in effect that the jumps of a Poisson process are as randomly distributed as possible.

Theorem 2.4.5. *Let $(X_t)_{t \geq 0}$ be a Poisson process. Then, conditional on $(X_t)_{t \geq 0}$ having exactly one jump in the interval $[s, s+t]$, the time at which that jump occurs is uniformly distributed on $[s, s+t]$.*

Proof. We shall use the finite-dimensional distribution definition. By stationarity of increments, it suffices to consider the case $s = 0$. Then, for $0 \leq u \leq t$,

$$\mathbb{P}(J_1 \leq u \mid X_t = 1) = \mathbb{P}(J_1 \leq u \text{ and } X_t = 1)/\mathbb{P}(X_t = 1)$$
$$= \mathbb{P}(X_u = 1 \text{ and } X_t - X_u = 0)/\mathbb{P}(X_t = 1)$$
$$= \lambda u e^{-\lambda u} e^{-\lambda(t-u)}/(\lambda t e^{-\lambda t}) = u/t. \qquad \square$$

Theorem 2.4.6. *Let $(X_t)_{t \geq 0}$ be a Poisson process. Then, conditional on the event $\{X_t = n\}$, the jump times $J_1, \ldots, J_n$ have joint density function*

$$f(t_1, \ldots, t_n) = n! t^{-n} 1_{\{0 \leq t_1 \leq \ldots \leq t_n \leq t\}}.$$

Thus, conditional on $\{X_t = n\}$, the jump times $J_1, \ldots, J_n$ have the same distribution as an ordered sample of size n from the uniform distribution on $[0, t]$.

Proof. The holding times $S_1, \ldots, S_{n+1}$ have joint density function

$$\lambda^{n+1} e^{-\lambda(s_1 + \ldots + s_{n+1})} 1_{\{s_1, \ldots, s_{n+1} \geq 0\}}$$

so the jump times $J_1, \ldots, J_{n+1}$ have joint density function

$$\lambda^{n+1} e^{-\lambda t_{n+1}} 1_{\{0 \leq t_1 \leq \ldots \leq t_{n+1}\}}.$$

So for $A \subseteq \mathbb{R}^n$ we have

$$\mathbb{P}\big((J_1, \ldots, J_n) \in A \text{ and } X_t = n\big) = \mathbb{P}\big((J_1, \ldots, J_n) \in A \text{ and } J_n \leq t < J_{n+1}\big)$$
$$= e^{-\lambda t} \lambda^n \int_{(t_1, \ldots, t_n) \in A} 1_{\{0 \leq t_1 \leq \ldots \leq t_n \leq t\}} dt_1 \ldots dt_n$$

and since $\mathbb{P}(X_t = n) = e^{-\lambda t}(\lambda t)^n/n!$ we obtain

$$\mathbb{P}\big((J_1, \ldots, J_n) \in A \mid X_t = n\big) = \int_A f(t_1, \ldots, t_n) dt_1 \ldots dt_n$$

as required. $\square$

We finish with a simple example typical of many problems making use of a range of properties of the Poisson process.

Example 2.4.7

Robins and blackbirds make brief visits to my birdtable. The probability that in any small interval of duration h a robin will arrive is found to be $\rho h + o(h)$, whereas the corresponding probability for blackbirds is $\beta h + o(h)$. What is the probability that the first two birds I see are both robins? What is the distribution of the total number of birds seen in time t? Given that this number is n, what is the distribution of the number of blackbirds seen in time t?

By the infinitesimal characterization, the number of robins seen by time t is a Poisson process $(R_t)_{t \geq 0}$ of rate ρ, and the number of blackbirds is a Poisson process $(B_t)_{t \geq 0}$ of rate β. The times spent waiting for the first robin or blackbird are independent exponential random variables S_1 and T_1 of parameters ρ and β respectively. So a robin arrives first with probability $\rho/(\rho + \beta)$ and, by the memoryless property of T_1, the probability that the first two birds are robins is $\rho^2/(\rho + \beta)^2$. By Theorem 2.4.4 the total number of birds seen in an interval of duration t has Poisson distribution of parameter $(\rho + \beta)t$. Finally

$$\mathbb{P}(B_t = k \mid R_t + B_t = n) = \mathbb{P}(B_t = k \text{ and } R_t = n-k)/\mathbb{P}(R_t + B_t = n)$$

$$= \left(\frac{e^{-\beta}\beta^k}{k!}\right)\left(\frac{e^{-\rho}\rho^{n-k}}{(n-k)!}\right) \Big/ \left(\frac{e^{-(\rho+\beta)}(\rho+\beta)^n}{n!}\right)$$

$$= \binom{n}{k}\left(\frac{\beta}{\rho+\beta}\right)^k\left(\frac{\rho}{\rho+\beta}\right)^{n-k}$$

so if n birds are seen in time t, then the distribution of the number of blackbirds is binomial of parameters n and $\beta/(\rho + \beta)$.

Exercises

2.4.1 State the transition probability definition of a Poisson process. Show directly from this definition that the first jump time J_1 of a Poisson process of rate λ is exponential of parameter λ.

Show also (from the same definition and without assuming the strong Markov property) that

$$\mathbb{P}(t_1 < J_1 \leq t_2 < J_2) = e^{-\lambda t_1}\lambda(t_2 - t_1)e^{-\lambda(t_2 - t_1)}$$

and hence that $J_2 - J_1$ is also exponential of parameter λ and independent of J_1.

2.4.2 Show directly from the infinitesimal definition that the first jump time J_1 of a Poisson process of rate λ has exponential distribution of parameter λ.

2.4.3 Arrivals of the Number 1 bus form a Poisson process of rate one bus per hour, and arrivals of the Number 7 bus form an independent Poisson process of rate seven buses per hour.

(a) What is the probability that exactly three buses pass by in one hour?

(b) What is the probability that exactly three Number 7 buses pass by while I am waiting for a Number 1?

(c) When the maintenance depot goes on strike half the buses break down before they reach my stop. What, then, is the probability that I wait for 30 minutes without seeing a single bus?

2.4.4 A radioactive source emits particles in a Poisson process of rate λ. The particles are each emitted in an independent random direction. A Geiger counter placed near the source records a fraction p of the particles emitted. What is the distribution of the number of particles recorded in time t?

2.4.5 A pedestrian wishes to cross a single lane of fast-moving traffic. Suppose the number of vehicles that have passed by time t is a Poisson process of rate λ, and suppose it takes time a to walk across the lane. Assuming that the pedestrian can foresee correctly the times at which vehicles will pass by, how long on average does it take to cross over safely? [*Consider the time at which the first car passes.*]

How long on average does it take to cross two similar lanes (a) when one must walk straight across (assuming that the pedestrian will not cross if, at any time whilst crossing, a car would pass in either direction), (b) when an island in the middle of the road makes it safe to stop half-way?

2.5 Birth processes

A birth process is a generalization of a Poisson process in which the parameter λ is allowed to depend on the current state of the process. The data for a birth process consist of *birth rates* $0 \leq q_j < \infty$, where $j = 0, 1, 2, \ldots$. We begin with a definition in terms of jump chain and holding times. A minimal right-continuous process $(X_t)_{t \geq 0}$ with values in $\{0, 1, 2, \ldots\} \cup \{\infty\}$ is a *birth process of rates* $(q_j : j \geq 0)$ if, conditional on $X_0 = i$, its holding times $S_1, S_2, \ldots$ are independent exponential random variables of parameters $q_i, q_{i+1}, \ldots$, respectively, and its jump chain is given by $Y_n = i + n$.

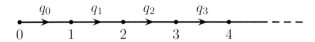

The flow diagram is shown above and the Q-matrix is given by:

$$Q = \begin{pmatrix} -q_0 & q_0 & & & \\ & -q_1 & q_1 & & \\ & & -q_2 & q_2 & \\ & & & \ddots & \ddots \end{pmatrix}.$$

Example 2.5.1 (Simple birth process)

Consider a population in which each individual gives birth after an exponential time of parameter λ, all independently. If i individuals are present then the first birth will occur after an exponential time of parameter $i\lambda$. Then we have $i+1$ individuals and, by the memoryless property, the process begins afresh. Thus the size of the population performs a birth process with rates $q_i = i\lambda$. Let X_t denote the number of individuals at time t and suppose $X_0 = 1$. Write T for the time of the first birth. Then

$$\mathbb{E}(X_t) = \mathbb{E}(X_t 1_{T \le t}) + \mathbb{E}(X_t 1_{T > t})$$
$$= \int_0^t \lambda e^{-\lambda s} \mathbb{E}(X_t \mid T = s) ds + e^{-\lambda t}.$$

Put $\mu(t) = \mathbb{E}(X_t)$, then $\mathbb{E}(X_t \mid T = s) = 2\mu(t-s)$, so

$$\mu(t) = \int_0^t 2\lambda e^{-\lambda s} \mu(t-s) ds + e^{-\lambda t}$$

and setting $r = t - s$

$$e^{\lambda t} \mu(t) = 2\lambda \int_0^t e^{\lambda r} \mu(r) dr + 1.$$

By differentiating we obtain

$$\mu'(t) = \lambda \mu(t)$$

so the mean population size grows exponentially:

$$\mathbb{E}(X_t) = e^{\lambda t}.$$

2.5 Birth processes

Much of the theory associated with the Poisson process goes through for birth processes with little change, except that some calculations can no longer be made so explicitly. The most interesting new phenomenon present in birth processes is the possibility of explosion. For certain choices of birth rates, a typical path will make infinitely many jumps in a finite time, as shown in the diagram. The convention of setting the process to equal ∞ after explosion is particularly appropriate for birth processes!

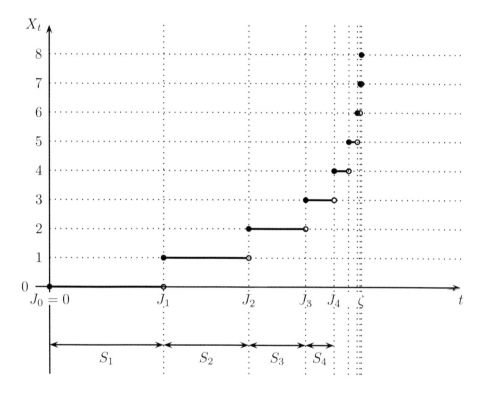

In fact, Theorem 2.3.2 tells us exactly when explosion will occur.

Theorem 2.5.2. *Let $(X_t)_{t\geq 0}$ be a birth process of rates $(q_j : j \geq 0)$, starting from 0.*

(i) *If* $\sum_{j=0}^{\infty} \dfrac{1}{q_j} < \infty$, *then* $\mathbb{P}(\zeta < \infty) = 1$.

(ii) *If* $\sum_{j=0}^{\infty} \dfrac{1}{q_j} = \infty$, *then* $\mathbb{P}(\zeta = \infty) = 1$.

Proof. Apply Theorem 2.3.2 to the sequence of holding times $S_1, S_2, \ldots$. □

The proof of the Markov property for the Poisson process is easily adapted to give the following generalization.

Theorem 2.5.3 (Markov property). *Let $(X_t)_{t \geq 0}$ be a birth process of rates $(q_j : j \geq 0)$. Then, conditional on $X_s = i$, $(X_{s+t})_{t \geq 0}$ is a birth process of rates $(q_j : j \geq 0)$ starting from i and independent of $(X_r : r \leq s)$.*

We shall shortly prove a theorem on birth processes which generalizes the key theorem on Poisson processes. First we must see what will replace the Poisson probabilities. In Theorem 2.4.3 these arose as the unique solution of a system of differential equations, which we showed were essentially the forward equations. Now we can still write down the forward equation

$$P'(t) = P(t)Q, \quad P(0) = I.$$

or, in components

$$p'_{i0}(t) = -p_{i0}(t)q_0, \quad p_{i0}(0) = \delta_{i0}$$

and, for $j = 1, 2, \ldots$

$$p'_{ij}(t) = p_{i,j-1}(t)q_{j-1} - p_{ij}(t)q_j, \quad p_{ij}(0) = \delta_{ij}.$$

Moreover, these equations still have a unique solution; it is just not as explicit as before. For we must have

$$p_{i0}(t) = \delta_{i0} e^{-q_0 t}$$

which can be substituted in the equation

$$p'_{i1}(t) = p_{i0}(t)q_0 - p_{i1}(t)q_1, \quad p_{i1}(0) = \delta_{i1}$$

and this equation solved to give

$$p_{i1}(t) = \delta_{i1} e^{-q_1 t} + \delta_{i0} \int_0^t q_0 e^{-q_0 s} e^{-q_1(t-s)} ds.$$

Now we can substitute for $p_{i1}(t)$ in the next equation up the hierarchy and find an explicit expression for $p_{i2}(t)$, and so on.

Theorem 2.5.4. *Let $(X_t)_{t \geq 0}$ be an increasing, right-continuous process with values in $\{0, 1, 2, \ldots\} \cup \{\infty\}$. Let $0 \leq q_j < \infty$ for all $j \geq 0$. Then the following three conditions are equivalent:*

(a) *(jump chain/holding time definition) conditional on $X_0 = i$, the holding times $S_1, S_2, \ldots$ are independent exponential random variables of parameters $q_i, q_{i+1}, \ldots$ respectively and the jump chain is given by $Y_n = i + n$ for all n;*

(b) (infinitesimal definition) for all $t, h \geq 0$, conditional on $X_t = i$, X_{t+h} is independent of $(X_s : s \leq t)$ and, as $h \downarrow 0$, uniformly in t,

$$\mathbb{P}(X_{t+h} = i \mid X_t = i) = 1 - q_i h + o(h),$$
$$\mathbb{P}(X_{t+h} = i+1 \mid X_t = i) = q_i h + o(h);$$

(c) (transition probability definition) for all $n = 0, 1, 2, \ldots$, all times $0 \leq t_0 \leq \ldots \leq t_{n+1}$ and all states $i_0, \ldots, i_{n+1}$

$$\mathbb{P}(X_{t_{n+1}} = i_{n+1} \mid X_{t_0} = i_0, \ldots, X_{t_n} = i_n) = p_{i_n i_{n+1}}(t_{n+1} - t_n)$$

where $(p_{ij}(t) : i, j = 0, 1, 2, \ldots)$ is the unique solution of the forward equations.

If $(X_t)_{t \geq 0}$ satisfies any of these conditions then it is called a *birth process* of rates $(q_j : j \geq 0)$.

Proof. (a) $\Rightarrow$ (b) If (a) holds, then, by the Markov property for any $t, h \geq 0$, conditional on $X_t = i$, X_{t+h} is independent of $(X_s : s \leq t)$ and, as $h \downarrow 0$, uniformly in t,

$$\mathbb{P}(X_{t+h} \geq i+1 \mid X_t = i) = \mathbb{P}(X_h \geq i+1 \mid X_0 = i)$$
$$= \mathbb{P}(J_1 \leq h \mid X_0 = i) = 1 - e^{-q_i h} = q_i h + o(h),$$

and

$$\mathbb{P}(X_{t+h} \geq i+2 \mid X_t = i) = \mathbb{P}(X_h \geq i+2 \mid X_0 = i)$$
$$= \mathbb{P}(J_2 \leq h \mid X_0 = i) \leq \mathbb{P}(S_1 \leq h \text{ and } S_2 \leq h \mid X_0 = i)$$
$$= (1 - e^{-q_i h})(1 - e^{-q_{i+1} h}) = o(h),$$

which implies (b).

(b) $\Rightarrow$ (c) If (b) holds, then certainly for $k = i+2, i+3, \ldots$

$$\mathbb{P}(X_{t+h} = k \mid X_t = i) = o(h) \quad \text{as } h \downarrow 0, \text{ uniformly in } t.$$

Set $p_{ij}(t) = \mathbb{P}(X_t = j \mid X_0 = i)$. Then, for $j = 1, 2, \ldots$

$$p_{ij}(t+h) = \mathbb{P}(X_{t+h} = j \mid X_0 = i)$$
$$= \sum_{k=i}^{j} \mathbb{P}(X_t = k \mid X_0 = i)\mathbb{P}(X_{t+h} = j \mid X_t = k)$$
$$= p_{ij}(t)(1 - q_j h + o(h)) + p_{i,j-1}(t)(q_{j-1} h + o(h)) + o(h)$$

so
$$\frac{p_{ij}(t+h) - p_{ij}(t)}{h} = p_{i,j-1}(t)q_{j-1} - p_{ij}(t)q_j + O(h).$$

As in the proof of Theorem 2.4.3, we can deduce that $p_{ij}(t)$ is differentiable and satisfies the differential equation

$$p'_{ij}(t) = p_{i,j-1}(t)q_{j-1} - p_{ij}(t)q_j.$$

By a simpler argument we also find

$$p'_{i0}(t) = -p_{i0}(t)q_0.$$

Thus $(p_{ij}(t) : i,j = 0,1,2,\dots)$ must be the unique solution to the forward equations. If $(X_t)_{t\geq 0}$ satisfies (b), then certainly

$$\mathbb{P}(X_{t_{n+1}} = i_{n+1} \mid X_0 = i_0, \dots, X_{t_n} = i_n) = \mathbb{P}(X_{t_{n+1}} = i_{n+1} \mid X_{t_n} = i_n)$$

but also $(X_{t_n+t})_{t\geq 0}$ satisfies (b), so

$$\mathbb{P}(X_{t_{n+1}} = i_{n+1} \mid X_{t_n} = i_n) = p_{i_n i_{n+1}}(t_{n+1} - t_n)$$

by uniqueness for the forward equations. Hence $(X_t)_{t\geq 0}$ satisfies (c).

(c) $\Rightarrow$ (a) See the proof of Theorem 2.4.3. □

Exercise

2.5.1 Each bacterium in a colony splits into two identical bacteria after an exponential time of parameter λ, which then split in the same way but independently. Let X_t denote the size of the colony at time t, and suppose $X_0 = 1$. Show that the probability generating function $\phi(t) = \mathbb{E}(z^{X_t})$ satisfies

$$\phi(t) = ze^{-\lambda t} + \int_0^t \lambda e^{-\lambda s} \phi(t-s)^2 ds.$$

Make a change of variables $u = t - s$ in the integral and deduce that $d\phi/dt = \lambda\phi(\phi - 1)$. Hence deduce that, for $q = 1 - e^{-\lambda t}$ and $n = 1, 2, \dots$

$$\mathbb{P}(X_t = n) = q^{n-1}(1-q).$$

2.6 Jump chain and holding times

This section begins the theory of continuous-time Markov chains proper, which will occupy the remainder of this chapter and the whole of the next. The approach we have chosen is to introduce continuous-time chains in terms of the joint distribution of their jump chain and holding times. This provides the most direct mathematical description. It also makes possible a number of constructive realizations of a given Markov chain, which we shall describe, and which underlie many applications.

Let I be a countable set. The basic data for a continuous-time Markov chain on I are given in the form of a Q-matrix. Recall that a Q-matrix on I is any matrix $Q = (q_{ij} : i, j \in I)$ which satisfies the following conditions:

(i) $0 \leq -q_{ii} < \infty$ for all i;
(ii) $q_{ij} \geq 0$ for all $i \neq j$;
(iii) $\sum_{j \in I} q_{ij} = 0$ for all i.

We will sometimes find it convenient to write q_i or $q(i)$ as an alternative notation for $-q_{ii}$.

We are going to describe a simple procedure for obtaining from a Q-matrix Q a stochastic matrix Π. The *jump matrix* $\Pi = (\pi_{ij} : i, j \in I)$ of Q is defined by

$$\pi_{ij} = \begin{cases} q_{ij}/q_i & \text{if } j \neq i \text{ and } q_i \neq 0 \\ 0 & \text{if } j \neq i \text{ and } q_i = 0, \end{cases}$$
$$\pi_{ii} = \begin{cases} 0 & \text{if } q_i \neq 0 \\ 1 & \text{if } q_i = 0. \end{cases}$$

This procedure is best thought of row by row. For each $i \in I$ we take, where possible, the off-diagonal entries in the ith row of Q and scale them so they add up to 1, putting a 0 on the diagonal. This is only impossible when the off-diagonal entries are all 0, then we leave them alone and put a 1 on the diagonal. As you will see in the following example, the associated diagram transforms into a discrete-time Markov chain diagram simply by rescaling all the numbers on any arrows leaving a state so they add up to 1.

Example 2.6.1

The Q-matrix

$$Q = \begin{pmatrix} -2 & 1 & 1 \\ 1 & -1 & 0 \\ 2 & 1 & -3 \end{pmatrix}$$

has diagram:

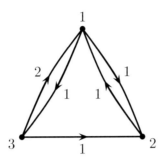

The jump matrix Π of Q is given by

$$\Pi = \begin{pmatrix} 0 & 1/2 & 1/2 \\ 1 & 0 & 0 \\ 2/3 & 1/3 & 0 \end{pmatrix}$$

and has diagram:

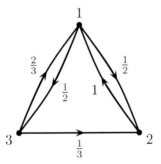

Here is the definition of a continuous-time Markov chain in terms of its jump chain and holding times. Recall that a minimal process is one which is set equal to ∞ after any explosion – see Section 2.2. A minimal right-continuous process $(X_t)_{t\geq 0}$ on I is a *Markov chain with initial distribution* λ *and generator matrix* Q if its jump chain $(Y_n)_{n\geq 0}$ is discrete-time Markov(λ, Π) and if for each $n \geq 1$, conditional on $Y_0, \ldots, Y_{n-1}$, its holding times $S_1, \ldots, S_n$ are independent exponential random variables of parameters $q(Y_0), \ldots, q(Y_{n-1})$ respectively. We say $(X_t)_{t\geq 0}$ is *Markov*(λ, Q) for short. We can construct such a process as follows: let $(Y_n)_{n\geq 0}$ be discrete-time Markov(λ, Π) and let $T_1, T_2, \ldots$ be independent exponential random

2.6 Jump chain and holding times

variables of parameter 1, independent of $(Y_n)_{n\geq 0}$. Set $S_n = T_n/q(Y_{n-1})$, $J_n = S_1 + \ldots + S_n$ and

$$X_t = \begin{cases} Y_n & \text{if } J_n \leq t < J_{n+1} \text{ for some } n \\ \infty & \text{otherwise.} \end{cases}$$

Then $(X_t)_{t\geq 0}$ has the required properties.

We shall now describe two further constructions. You will need to understand these constructions in order to identify processes in applications which can be modelled as Markov chains. Both constructions make direct use of the entries in the Q-matrix, rather than proceeding first via the jump matrix. Here is the second construction.

We begin with an initial state $X_0 = Y_0$ with distribution λ, and with an array $(T_n^j : n \geq 1, j \in I)$ of independent exponential random variables of parameter 1. Then, inductively for $n = 0, 1, 2, \ldots$, if $Y_n = i$ we set

$$S_{n+1}^j = T_{n+1}^j/q_{ij}, \quad \text{for } j \neq i,$$

$$S_{n+1} = \inf_{j \neq i} S_{n+1}^j,$$

$$Y_{n+1} = \begin{cases} j & \text{if } S_{n+1}^j = S_{n+1} < \infty \\ i & \text{if } S_{n+1} = \infty. \end{cases}$$

Then, conditional on $Y_n = i$, the random variables S_{n+1}^j are independent exponentials of parameter q_{ij} for all $j \neq i$. So, conditional on $Y_n = i$, by Theorem 2.3.3, S_{n+1} is exponential of parameter $q_i = \sum_{j \neq i} q_{ij}$, Y_{n+1} has distribution $(\pi_{ij} : j \in I)$, and S_{n+1} and Y_{n+1} are independent, and independent of $Y_0, \ldots, Y_n$ and $S_1, \ldots, S_n$, as required. This construction shows why we call q_i the *rate of leaving i* and q_{ij} the *rate of going from i to j*.

Our third and final construction of a Markov chain with generator matrix Q and initial distribution λ is based on the Poisson process. Imagine the state-space I as a labyrinth of chambers and passages, each passage shut off by a single door which opens briefly from time to time to allow you through in one direction only. Suppose the door giving access to chamber j from chamber i opens at the jump times of a Poisson process of rate q_{ij} and you take every chance to move that you can, then you will perform a Markov chain with Q-matrix Q. In more mathematical terms, we begin with an initial state $X_0 = Y_0$ with distribution λ, and with a family of independent Poisson processes $\{(N_t^{ij})_{t\geq 0} : i, j \in I, i \neq j\}$, $(N_t^{ij})_{t\geq 0}$ having rate q_{ij}. Then set $J_0 = 0$ and define inductively for $n = 0, 1, 2, \ldots$

$$J_{n+1} = \inf\{t > J_n : N_t^{Y_n j} \neq N_{J_n}^{Y_n j} \text{ for some } j \neq Y_n\}$$

$$Y_{n+1} = \begin{cases} j & \text{if } J_{n+1} < \infty \text{ and } N_{J_{n+1}}^{Y_n j} \neq N_{J_n}^{Y_n j} \\ i & \text{if } J_{n+1} = \infty. \end{cases}$$

The first jump time of $(N_t^{ij})_{t \geq 0}$ is exponential of parameter q_{ij}. So, by Theorem 2.3.3, conditional on $Y_0 = i$, J_1 is exponential of parameter $q_i = \sum_{j \neq i} q_{ij}$, Y_1 has distribution $(\pi_{ij} : j \in I)$, and J_1 and Y_1 are independent.

Now suppose T is a stopping time of $(X_t)_{t \geq 0}$. If we condition on X_0 and on the processes $(N_t^{kl})_{t \geq 0}$ for $(k,l) \neq (i,j)$, which are independent of N_t^{ij}, then $\{T \leq t\}$ depends only on $(N_s^{ij} : s \leq t)$. So, by the strong Markov property of the Poisson process $\tilde{N}_t^{ij} := N_{T+t}^{ij} - N_T^{ij}$ is a Poisson process of rate q_{ij} independent of $(N_s^{ij} : s \leq T)$, and independent of X_0 and $(N_t^{kl})_{t \geq 0}$ for $(k,l) \neq (i,j)$. Hence, conditional on $T < \infty$ and $X_T = i$, $(X_{T+t})_{t \geq 0}$ has the same distribution as $(X_t)_{t \geq 0}$ and is independent of $(X_s : s \leq T)$. In particular, we can take $T = J_n$ to see that, conditional on $J_n < \infty$ and $Y_n = i$, S_{n+1} is exponential of parameter q_i, Y_{n+1} has distribution $(\pi_{ij} : j \in I)$, and S_{n+1} and Y_{n+1} are independent, and independent of $Y_0, \ldots, Y_n$ and $S_1, \ldots, S_n$. Hence $(X_t)_{t \geq 0}$ is Markov(λ, Q) and, moreover, $(X_t)_{t \geq 0}$ has the strong Markov property. The conditioning on which this argument relies requires some further justification, especially when the state-space is infinite, so we shall not rely on this third construction in the development of the theory.

2.7 Explosion

We saw in the special case of birth processes that, although each holding time is strictly positive, one can run through a sequence of states with shorter and shorter holding times and end up taking infinitely many jumps in a finite time. This phenomenon is called explosion. Recall the notation of Section 2.2: for a process with jump times $J_0, J_1, J_2, \ldots$ and holding times $S_1, S_2, \ldots$, the explosion time ζ is given by

$$\zeta = \sup_n J_n = \sum_{n=1}^{\infty} S_n.$$

Theorem 2.7.1. *Let $(X_t)_{t \geq 0}$ be Markov(λ, Q). Then $(X_t)_{t \geq 0}$ does not explode if any one of the following conditions holds:*

(i) *I is finite;*
(ii) *$\sup_{i \in I} q_i < \infty$;*
(iii) *$X_0 = i$, and i is recurrent for the jump chain.*

Proof. Set $T_n = q(Y_{n-1}) S_n$, then $T_1, T_2, \ldots$ are independent $E(1)$ and independent of $(Y_n)_{n \geq 0}$. In cases (i) and (ii), $q = \sup_i q_i < \infty$ and

$$q\zeta \geq \sum_{n=1}^{\infty} T_n = \infty$$

with probability 1. In case (iii), we know that $(Y_n)_{n\geq 0}$ visits i infinitely often, at times $N_1, N_2, \ldots$, say. Then

$$q_i \zeta \geq \sum_{m=1}^{\infty} T_{N_m+1} = \infty$$

with probability 1. □

We say that a Q-matrix Q is *explosive* if, for the associated Markov chain

$$\mathbb{P}_i(\zeta < \infty) > 0 \quad \text{for some } i \in I.$$

Otherwise Q is *non-explosive*. Here as in Chapter 1 we denote by $\mathbb{P}_i$ the conditional probability $\mathbb{P}_i(A) = \mathbb{P}(A|X_0 = i)$. It is a simple consequence of the Markov property for $(Y_n)_{n\geq 0}$ that under $\mathbb{P}_i$ the process $(X_t)_{t\geq 0}$ is Markov(δ_i, Q). The result just proved gives simple conditions for non-explosion and covers many cases of interest. As a corollary to the next result we shall obtain necessary and sufficient conditions for Q to be explosive, but these are not as easy to apply as Theorem 2.7.1.

Theorem 2.7.2. *Let $(X_t)_{t\geq 0}$ be a continuous-time Markov chain with generator matrix Q and write ζ for the explosion time of $(X_t)_{t\geq 0}$. Fix $\theta > 0$ and set $z_i = \mathbb{E}_i(e^{-\theta \zeta})$. Then $z = (z_i : i \in I)$ satisfies:*

  (i) $|z_i| \leq 1$ for all i;
  (ii) $Qz = \theta z$.

Moreover, if $\tilde{z}$ also satisfies (i) and (ii), then $\tilde{z}_i \leq z_i$ for all i.

Proof. Condition on $X_0 = i$. The time and place of the first jump are independent, J_1 is $E(q_i)$ and

$$\mathbb{P}_i(X_{J_1} = k) = \pi_{ik}.$$

Moreover, by the Markov property of the jump chain at time $n = 1$, conditional on $X_{J_1} = k$, $(X_{J_1+t})_{t\geq 0}$ is Markov(δ_k, Q) and independent of J_1. So

$$\mathbb{E}_i(e^{-\theta \zeta} \mid X_{J_1} = k) = \mathbb{E}_i(e^{-\theta J_1} e^{-\theta \sum_{n=2}^{\infty} S_n} \mid X_{J_1} = k)$$
$$= \int_0^{\infty} e^{-\theta t} q_i e^{-q_i t} dt \, \mathbb{E}_k(e^{-\theta \zeta}) = \frac{q_i z_k}{q_i + \theta}$$

and

$$z_i = \sum_{k \neq i} \mathbb{P}_i(X_{J_1} = k) \mathbb{E}_i(e^{-\theta \zeta} \mid X_{J_1} = k) = \sum_{k \neq i} \frac{q_i \pi_{ik} z_k}{q_i + \theta}.$$

Recall that $q_i = -q_{ii}$ and $q_i \pi_{ik} = q_{ik}$. Then

$$(\theta - q_{ii})z_i = \sum_{k \neq i} q_{ik} z_k$$

so

$$\theta z_i = \sum_{k \in I} q_{ik} z_k$$

and so z satisfies (i) and (ii). Note that the same argument also shows that

$$\mathbb{E}_i(e^{-\theta J_{n+1}}) = \sum_{k \neq i} \frac{q_i \pi_{ik}}{q_i + \theta} \mathbb{E}_k(e^{-\theta J_n}).$$

Suppose that $\tilde{z}$ also satisfies (i) and (ii), then, in particular

$$\tilde{z}_i \leq 1 = \mathbb{E}_i(e^{-\theta J_0})$$

for all i. Suppose inductively that

$$\tilde{z}_i \leq \mathbb{E}_i(e^{-\theta J_n})$$

then, since $\tilde{z}$ satisfies (ii)

$$\tilde{z}_i = \sum_{k \neq i} \frac{q_i \pi_{ik}}{q_i + \theta} \tilde{z}_k \leq \sum_{k \neq i} \frac{q_i \pi_{ik}}{q_i + \theta} \mathbb{E}_i(e^{-\theta J_n}) = \mathbb{E}_i(e^{-\theta J_{n+1}}).$$

Hence $\tilde{z}_i \leq \mathbb{E}_i(e^{-\theta J_n})$ for all n. By monotone convergence

$$\mathbb{E}_i(e^{-\theta J_n}) \to \mathbb{E}_i(e^{-\theta \zeta})$$

as $n \to \infty$, so $\tilde{z}_i \leq z_i$ for all i. $\square$

Corollary 2.7.3. *For each $\theta > 0$ the following are equivalent:*
  (a) *Q is non-explosive;*
  (b) *$Qz = \theta z$ and $|z_i| \leq 1$ for all i imply $z = 0$.*

Proof. If (a) holds then $\mathbb{P}_i(\zeta = \infty) = 1$ so $\mathbb{E}_i(e^{-\theta \zeta}) = 0$. By the theorem, $Qz = \theta z$ and $|z| \leq 1$ imply $z_i \leq \mathbb{E}_i(e^{-\theta \zeta})$, hence $z \leq 0$, by symmetry $z \geq 0$, and hence (b) holds. On the other hand, if (b) holds, then by the theorem $\mathbb{E}_i(e^{-\theta \zeta}) = 0$ for all i, so $\mathbb{P}_i(\zeta = \infty) = 1$ and (a) holds. $\square$

Exercise

2.7.1 Let $(X_t)_{t\geq 0}$ be a Markov chain on the integers with transition rates
$$q_{i,i+1} = \lambda q_i, \quad q_{i,i-1} = \mu q_i$$
and $q_{ij} = 0$ if $|j - i| \geq 2$, where $\lambda + \mu = 1$ and $q_i > 0$ for all i. Find for all integers i:

(a) the probability, starting from 0, that X_t hits i;

(b) the expected total time spent in state i, starting from 0.

In the case where $\mu = 0$, write down a necessary and sufficient condition for $(X_t)_{t\geq 0}$ to be explosive. Why is this condition necessary for $(X_t)_{t\geq 0}$ to be explosive for all $\mu \in [0, 1/2)$?

Show that, in general, $(X_t)_{t\geq 0}$ is non-explosive if and only if one of the following conditions holds:

(i) $\lambda = \mu$;
(ii) $\lambda > \mu$ and $\sum_{i=1}^{\infty} 1/q_i = \infty$;
(iii) $\lambda < \mu$ and $\sum_{i=1}^{\infty} 1/q_{-i} = \infty$.

2.8 Forward and backward equations

Although the definition of a continuous-time Markov chain in terms of its jump chain and holding times provides a clear picture of the process, it does not answer some basic questions. For example, we might wish to calculate $\mathbb{P}_i(X_t = j)$. In this section we shall obtain two more ways of characterizing a continuous-time Markov chain, which will in particular give us a means to find $\mathbb{P}_i(X_t = j)$. As for Poisson processes and birth processes, the first step is to deduce the *Markov property* from the jump chain/holding time definition. In fact, we shall give the *strong* Markov property as this is a fundamental result and the proof is not much harder. However, the proof of both results really requires the precision of measure theory, so we have deferred it to Section 6.5. If you want to understand what happens, Theorem 2.4.1 on the Poisson process gives the main idea in a simpler context.

Recall that a random variable T with values in $[0, \infty]$ is a stopping time of $(X_t)_{t\geq 0}$ if for each $t \in [0, \infty)$ the event $\{T \leq t\}$ depends only on $(X_s : s \leq t)$.

Theorem 2.8.1 (Strong Markov property). Let $(X_t)_{t\geq 0}$ be Markov(λ, Q) and let T be a stopping time of $(X_t)_{t\geq 0}$. Then, conditional on $T < \infty$ and $X_T = i$, $(X_{T+t})_{t\geq 0}$ is Markov(δ_i, Q) and independent of $(X_s : s \leq T)$.

We come to the key result for continuous-time Markov chains. We shall present first a version for the case of finite state-space, where there is a simpler proof. In this case there are three alternative definitions, just as for the Poisson process.

Theorem 2.8.2. *Let $(X_t)_{t\geq 0}$ be a right-continuous process with values in a finite set I. Let Q be a Q-matrix on I with jump matrix Π. Then the following three conditions are equivalent:*

(a) *(jump chain/holding time definition) conditional on $X_0 = i$, the jump chain $(Y_n)_{n\geq 0}$ of $(X_t)_{t\geq 0}$ is discrete-time Markov(δ_i, Π) and for each $n \geq 1$, conditional on $Y_0, \ldots, Y_{n-1}$, the holding times $S_1, \ldots, S_n$ are independent exponential random variables of parameters $q(Y_0), \ldots, q(Y_{n-1})$ respectively;*

(b) *(infinitesimal definition) for all $t, h \geq 0$, conditional on $X_t = i$, X_{t+h} is independent of $(X_s : s \leq t)$ and, as $h \downarrow 0$, uniformly in t, for all j*

$$\mathbb{P}(X_{t+h} = j \mid X_t = i) = \delta_{ij} + q_{ij}h + o(h);$$

(c) *(transition probability definition) for all $n = 0, 1, 2, \ldots$, all times $0 \leq t_0 \leq t_1 \leq \ldots \leq t_{n+1}$ and all states $i_0, \ldots, i_{n+1}$*

$$\mathbb{P}(X_{t_{n+1}} = i_{n+1} \mid X_{t_0} = i_0, \ldots, X_{t_n} = i_n) = p_{i_n i_{n+1}}(t_{n+1} - t_n)$$

where $(p_{ij}(t) : i, j \in I, t \geq 0)$ is the solution of the forward equation

$$P'(t) = P(t)Q, \quad P(0) = I.$$

If $(X_t)_{t\geq 0}$ satisfies any of these conditions then it is called a *Markov chain with generator matrix Q*. We say that $(X_t)_{t\geq 0}$ is *Markov(λ, Q)* for short, where λ is the distribution of X_0.

Proof. (a) $\Rightarrow$ (b) Suppose (a) holds, then, as $h \downarrow 0$,

$$\mathbb{P}_i(X_h = i) \geq \mathbb{P}_i(J_1 > h) = e^{-q_i h} = 1 + q_{ii}h + o(h)$$

and for $j \neq i$ we have

$$\mathbb{P}_i(X_h = j) \geq \mathbb{P}_i(J_1 \leq h, Y_1 = j, S_2 > h)$$
$$= (1 - e^{-q_i h})\pi_{ij} e^{-q_j h} = q_{ij}h + o(h).$$

Thus for every state j there is an inequality

$$\mathbb{P}_i(X_h = j) \geq \delta_{ij} + q_{ij}h + o(h)$$

and by taking the finite sum over j we see that these must in fact be equalities. Then by the Markov property, for any $t, h \geq 0$, conditional on $X_t = i$, X_{t+h} is independent of $(X_s : s \leq t)$ and, as $h \downarrow 0$, uniformly in t

$$\mathbb{P}(X_{t+h} = j \mid X_t = i) = \mathbb{P}_i(X_h = j) = \delta_{ij} + q_{ij}h + o(h).$$

(b) $\Rightarrow$ (c) Set $p_{ij}(t) = \mathbb{P}_i(X_t = j) = \mathbb{P}(X_t = j \mid X_0 = i)$. If (b) holds, then for all $t, h \geq 0$, as $h \downarrow 0$, uniformly in t

$$p_{ij}(t+h) = \sum_{k \in I} \mathbb{P}_i(X_t = k)\mathbb{P}(X_{t+h} = j \mid X_t = k)$$
$$= \sum_{k \in I} p_{ik}(t)(\delta_{kj} + q_{kj}h + o(h)).$$

Since I is finite we have

$$\frac{p_{ij}(t+h) - p_{ij}(t)}{h} = \sum_{k \in I} p_{ik}(t) q_{kj} + O(h)$$

so, letting $h \downarrow 0$, we see that $p_{ij}(t)$ is differentiable on the right. Then by uniformity we can replace t by $t - h$ in the above and let $h \downarrow 0$ to see first that $p_{ij}(t)$ is continuous on the left, then differentiable on the left, hence differentiable, and satisfies the forward equations

$$p'_{ij}(t) = \sum_{k \in I} p_{ik}(t) q_{kj}, \quad p_{ij}(0) = \delta_{ij}.$$

Since I is finite, $p_{ij}(t)$ is then the unique solution by Theorem 2.1.1. Also, if (b) holds, then

$$\mathbb{P}(X_{t_{n+1}} = i_{n+1} \mid X_{t_0} = i_0, \ldots, X_{t_n} = i_n) = \mathbb{P}(X_{t_{n+1}} = i_{n+1} \mid X_{t_n} = i_n)$$

and, moreover, (b) holds for $(X_{t_n+t})_{t \geq 0}$ so, by the above argument,

$$\mathbb{P}(X_{t_{n+1}} = i_{n+1} \mid X_{t_n} = i_n) = p_{i_n i_{n+1}}(t_{n+1} - t_n),$$

proving (c).

(c) $\Rightarrow$ (a) See the proof of Theorem 2.4.3. □

We know from Theorem 2.1.1 that for I finite the forward and backward equations have the same solution. So in condition (c) of the result just proved we could replace the forward equation with the backward equation. Indeed, there is a slight variation of the argument from (b) to (c) which leads directly to the backward equation.

The deduction of (c) from (b) above can be seen as the matrix version of the following result: for $q \in \mathbb{R}$ we have
$$\left(1 + \frac{q}{n} + o\left(\frac{1}{n}\right)\right)^n \to e^q \quad \text{as } n \to \infty.$$
Suppose (b) holds and set
$$p_{ij}(t, t+h) = \mathbb{P}(X_{t+h} = j \mid X_t = i);$$
then $P(t, t+h) = (p_{ij}(t, t+h) : i, j \in I)$ satisfies
$$P(t, t+h) = I + Qh + o(h)$$
and
$$P(0, t) = P\left(0, \frac{t}{n}\right) P\left(\frac{t}{n}, \frac{2t}{n}\right) \ldots P\left(\frac{(n-1)t}{n}, t\right) = \left(I + \frac{tQ}{n} + o\left(\frac{1}{n}\right)\right)^n.$$
Some care is needed in making this precise, since the $o(h)$ terms, though uniform in t, are not *a priori* identical. On the other hand, in (c) we see that
$$P(0, t) = e^{tQ}.$$

We turn now to the case of infinite state-space. The backward equation may still be written in the form
$$P'(t) = QP(t), \quad P(0) = I$$
only now we have an infinite system of differential equations
$$p'_{ij}(t) = \sum_{k \in I} q_{ik} p_{kj}(t), \quad p_{ij}(0) = \delta_{ij}$$
and the results on matrix exponentials given in Section 2.1 no longer apply. A solution to the backward equation is any matrix $(p_{ij}(t) : i, j \in I)$ of differentiable functions satisfying this system of differential equations.

Theorem 2.8.3. *Let Q be a Q-matrix. Then the backward equation*
$$P'(t) = QP(t), \quad P(0) = I$$
has a minimal non-negative solution $(P(t) : t \geq 0)$. This solution forms a matrix semigroup
$$P(s)P(t) = P(s+t) \quad \text{for all } s, t \geq 0.$$

We shall prove this result by a probabilistic method in combination with Theorem 2.8.4. Note that if I is finite we must have $P(t) = e^{tQ}$ by Theorem 2.1.1. We call $(P(t) : t \geq 0)$ the *minimal non-negative semigroup* associated to Q, or simply the *semigroup* of Q, the qualifications *minimal* and *non-negative* being understood.

Here is the key result for Markov chains with infinite state-space. There are just two alternative definitions now as the infinitesimal characterization becomes problematic for infinite state-space.

Theorem 2.8.4. *Let $(X_t)_{t\geq 0}$ be a minimal right-continuous process with values in I. Let Q be a Q-matrix on I with jump matrix Π and semigroup $(P(t) : t \geq 0)$. Then the following conditions are equivalent:*

(a) *(jump chain/holding time definition) conditional on $X_0 = i$, the jump chain $(Y_n)_{n\geq 0}$ of $(X_t)_{t\geq 0}$ is discrete-time Markov(δ_i, Π) and for each $n \geq 1$, conditional on $Y_0, \ldots, Y_{n-1}$, the holding times $S_1, \ldots, S_n$ are independent exponential random variables of parameters $q(Y_0), \ldots, q(Y_{n-1})$ respectively;*

(b) *(transition probability definition) for all $n = 0, 1, 2, \ldots$, all times $0 \leq t_0 \leq t_1 \leq \ldots \leq t_{n+1}$ and all states $i_0, i_1, \ldots, i_{n+1}$*
$$\mathbb{P}(X_{t_{n+1}} = i_{n+1} \mid X_{t_0} = i_0, \ldots, X_{t_n} = i_n) = p_{i_n i_{n+1}}(t_{n+1} - t_n).$$

If $(X_t)_{t\geq 0}$ satisfies any of these conditions then it is called a Markov chain with generator matrix Q. We say that $(X_t)_{t\geq 0}$ is Markov(λ, Q) for short, where λ is the distribution of X_0.

Proof of Theorems 2.8.3 and 2.8.4. We know that there exists a process $(X_t)_{t\geq 0}$ satisfying (a). So let us *define* $P(t)$ by
$$p_{ij}(t) = \mathbb{P}_i(X_t = j).$$

Step 1. *We show that $P(t)$ satisfies the backward equation.*

Conditional on $X_0 = i$ we have $J_1 \sim E(q_i)$ and $X_{J_1} \sim (\pi_{ik} : k \in I)$. Then conditional on $J_1 = s$ and $X_{J_1} = k$ we have $(X_{s+t})_{t\geq 0} \sim$ Markov(δ_k, Q). So
$$\mathbb{P}_i(X_t = j, t < J_1) = e^{-q_i t}\delta_{ij}$$
and
$$\mathbb{P}_i(J_1 \leq t, X_{J_1} = k, X_t = j) = \int_0^t q_i e^{-q_i s}\pi_{ik} p_{kj}(t-s)\, ds.$$
Therefore
$$p_{ij}(t) = \mathbb{P}_i(X_t = j, t < J_1) + \sum_{k\neq i}\mathbb{P}_i(J_1 \leq t, X_{J_1} = k, X_t = j)$$
$$= e^{-q_i t}\delta_{ij} + \sum_{k\neq i}\int_0^t q_i e^{-q_i s}\pi_{ik} p_{kj}(t-s)\, ds. \tag{2.1}$$

Make a change of variable $u = t - s$ in each of the integrals, interchange sum and integral by monotone convergence and multiply by $e^{q_i t}$ to obtain
$$e^{q_i t}p_{ij}(t) = \delta_{ij} + \int_0^t \sum_{k\neq i} q_i e^{q_i u}\pi_{ik} p_{kj}(u)\, du. \tag{2.2}$$

This equation shows, firstly, that $p_{ij}(t)$ is continuous in t for all i,j. Secondly, the integrand is then a uniformly converging sum of continuous functions, hence continuous, and hence $p_{ij}(t)$ is differentiable in t and satisfies

$$e^{q_i t}(q_i p_{ij}(t) + p'_{ij}(t)) = \sum_{k \neq i} q_i e^{q_i t} \pi_{ik} p_{kj}(t).$$

Recall that $q_i = -q_{ii}$ and $q_{ik} = q_i \pi_{ik}$ for $k \neq i$. Then, on rearranging, we obtain

$$p'_{ij}(t) = \sum_{k \in I} q_{ik} p_{kj}(t) \tag{2.3}$$

so $P(t)$ satisfies the backward equation.

The integral equation (2.1) is called the *integral form of the backward equation*.

Step 2. *We show that if $\widetilde{P}(t)$ is another non-negative solution of the backward equation, then $P(t) \leq \widetilde{P}(t)$, hence $P(t)$ is the minimal non-negative solution.*

The argument used to prove (2.1) also shows that

$$\mathbb{P}_i(X_t = j, t < J_{n+1})$$
$$= e^{-q_i t} \delta_{ij} + \sum_{k \neq i} \int_0^t q_i e^{-q_i s} \pi_{ik} \mathbb{P}_k(X_{t-s} = j, t-s < J_n) ds. \tag{2.4}$$

On the other hand, if $\widetilde{P}(t)$ satisfies the backward equation, then, by reversing the steps from (2.1) to (2.3), it also satisfies the integral form:

$$\widetilde{p}_{ij}(t) = e^{-q_i t} \delta_{ij} + \sum_{k \neq i} \int_0^t q_i e^{-q_i s} \pi_{ik} \widetilde{p}_{kj}(t-s) ds. \tag{2.5}$$

If $\widetilde{P}(t) \geq 0$, then

$$\mathbb{P}_i(X_t = j, t < J_0) = 0 \leq \widetilde{p}_{ij}(t) \qquad \text{for all } i,j \text{ and } t.$$

Let us suppose inductively that

$$\mathbb{P}_i(X_t = j, t < J_n) \leq \widetilde{p}_{ij}(t) \qquad \text{for all } i,j \text{ and } t,$$

then by comparing (2.4) and (2.5) we have

$$\mathbb{P}_i(X_t = j, t < J_{n+1}) \leq \widetilde{p}_{ij}(t) \qquad \text{for all } i,j \text{ and } t,$$

and the induction proceeds. Hence

$$p_{ij}(t) = \lim_{n \to \infty} \mathbb{P}_i(X_t = j, t < J_n) \leq \tilde{p}_{ij}(t) \quad \text{for all } i,j \text{ and } t.$$

Step 3. Since $(X_t)_{t \geq 0}$ does not return from ∞ we have

$$p_{ij}(s+t) = \mathbb{P}_i(X_{s+t} = j) = \sum_{k \in I} \mathbb{P}_i(X_{s+t} = j \mid X_s = k)\mathbb{P}_i(X_s = k)$$

$$= \sum_{k \in I} \mathbb{P}_i(X_s = k)\mathbb{P}_k(X_t = j) = \sum_{k \in I} p_{ik}(s)p_{kj}(t)$$

by the Markov property. Hence $(P(t) : t \geq 0)$ is a matrix semigroup. This completes the proof of Theorem 2.8.3.

Step 4. Suppose, as we have throughout, that $(X_t)_{t \geq 0}$ satisfies (a). Then, by the Markov property

$$\mathbb{P}(X_{t_{n+1}} = i_{n+1} \mid X_{t_0} = i_0, \ldots, X_{t_n} = i_n)$$
$$= \mathbb{P}_{i_n}(X_{t_{n+1}-t_n} = i_{n+1}) = p_{i_n i_{n+1}}(t_{n+1} - t_n)$$

so $(X_t)_{t \geq 0}$ satisfies (b). We complete the proof of Theorem 2.8.4 by the usual argument that (b) must now imply (a) (see the proof of Theorem 2.4.3, (c) $\Rightarrow$ (a)). □

So far we have said nothing about the forward equation in the case of infinite state-space. Remember that the finite state-space results of Section 2.1 are no longer valid. The forward equation may still be written

$$P'(t) = P(t)Q, \quad P(0) = I,$$

now understood as an infinite system of differential equations

$$p'_{ij}(t) = \sum_{k \in I} p_{ik}(t)q_{kj}, \quad p_{ij}(0) = \delta_{ij}.$$

A solution is then any matrix $(p_{ij}(t) : i, j \in I)$ of differentiable functions satisfying this system of equations. We shall show that the semigroup $(P(t) : t \geq 0)$ of Q does satisfy the forward equations, by a probabilistic argument resembling Step 1 of the proof of Theorems 2.8.3 and 2.8.4. This time, instead of conditioning on the first event, we condition on the last event before time t. The argument is a little longer because there is no reverse-time Markov property to give the conditional distribution. We need the following time-reversal identity, a simple version of which was given in Theorem 2.3.4.

Lemma 2.8.5. *We have*

$$q_{i_n}\mathbb{P}(J_n \leq t < J_{n+1} \mid Y_0 = i_0, Y_1 = i_1, \ldots, Y_n = i_n)$$
$$= q_{i_0}\mathbb{P}(J_n \leq t < J_{n+1} \mid Y_0 = i_n, \ldots, Y_{n-1} = i_1, Y_n = i_0).$$

Proof. Conditional on $Y_0 = i_0, \ldots, Y_n = i_n$, the holding times $S_1, \ldots, S_{n+1}$ are independent with $S_k \sim E(q_{i_{k-1}})$. So the left-hand side is given by

$$\int_{\Delta(t)} q_{i_n} \exp\{-q_{i_n}(t - s_1 - \ldots - s_n)\} \prod_{k=1}^{n} q_{i_{k-1}} \exp\{-q_{i_{k-1}} s_k\} ds_k$$

where $\Delta(t) = \{(s_1, \ldots, s_n) : s_1 + \ldots + s_n \leq t \text{ and } s_1, \ldots, s_n \geq 0\}$. On making the substitutions $u_1 = t - s_1 - \ldots - s_n$ and $u_k = s_{n-k+2}$, for $k = 2, \ldots, n$, we obtain

$$q_{i_n}\mathbb{P}(J_n \leq t < J_{n+1} \mid Y_0 = i_0, \ldots, Y_n = i_n)$$
$$= \int_{\Delta(t)} q_{i_0} \exp\{-q_{i_0}(t - u_1 - \ldots - u_n)\} \prod_{k=1}^{n} q_{i_{n-k+1}} \exp\{-q_{i_{n-k+1}} u_k\} du_k$$
$$= q_{i_0}\mathbb{P}(J_n \leq t < J_{n+1} \mid Y_0 = i_n, \ldots, Y_{n-1} = i_1, Y_n = i_0). \qquad \square$$

Theorem 2.8.6. *The minimal non-negative solution $(P(t) : t \geq 0)$ of the backward equation is also the minimal non-negative solution of the forward equation*

$$P'(t) = P(t)Q, \quad P(0) = I.$$

Proof. Let $(X_t)_{t \geq 0}$ denote the minimal Markov chain with generator matrix Q. By Theorem 2.8.4

$$p_{ij}(t) = \mathbb{P}_i(X_t = j)$$
$$= \sum_{n=0}^{\infty} \sum_{k \neq j} \mathbb{P}_i(J_n \leq t < J_{n+1}, Y_{n-1} = k, Y_n = j).$$

Now by Lemma 2.8.5, for $n \geq 1$, we have

$$\mathbb{P}_i(J_n \leq t < J_{n+1} \mid Y_{n-1} = k, Y_n = j)$$
$$= (q_i/q_j)\mathbb{P}_j(J_n \leq t < J_{n+1} \mid Y_1 = k, Y_n = i)$$
$$= (q_i/q_j) \int_0^t q_j e^{-q_j s} \mathbb{P}_k(J_{n-1} \leq t - s < J_n \mid Y_{n-1} = i) ds$$
$$= q_i \int_0^t e^{-q_j s} (q_k/q_i) \mathbb{P}_i(J_{n-1} \leq t - s < J_n \mid Y_{n-1} = k) ds$$

where we have used the Markov property of $(Y_n)_{n\geq 0}$ for the second equality. Hence

$$p_{ij}(t) = \delta_{ij}e^{-q_i t} + \sum_{n=1}^{\infty}\sum_{k\neq j}\int_0^t \mathbb{P}_i(J_{n-1} \leq t-s < J_n \mid Y_{n-1}=k)$$
$$\times \mathbb{P}_i(Y_{n-1}=k, Y_n=j)q_k e^{-q_j s}ds$$
$$= \delta_{ij}e^{-q_i t} + \sum_{n=1}^{\infty}\sum_{k\neq j}\int_0^t \mathbb{P}_i(J_{n-1} \leq t-s < J_n, Y_{n-1}=k)q_k \pi_{kj} e^{-q_j s}ds$$
$$= \delta_{ij}e^{-q_i t} + \int_0^t \sum_{k\neq j} p_{ik}(t-s) q_{kj} e^{-q_j s} ds \qquad (2.6)$$

where we have used monotone convergence to interchange the sum and integral at the last step. This is the *integral form of the forward equation*. Now make a change of variable $u = t - s$ in the integral and multiply by $e^{q_j t}$ to obtain

$$p_{ij}(t)e^{q_j t} = \delta_{ij} + \int_0^t \sum_{k\neq j} p_{ik}(u) q_{kj} e^{q_j u} du. \qquad (2.7)$$

We know by equation (2.2) that $e^{q_i t}p_{ik}(t)$ is *increasing* for all i, k. Hence either

$$\sum_{k\neq j} p_{ik}(u)q_{kj} \quad \text{converges uniformly for} \quad u \in [0, t]$$

or

$$\sum_{k\neq j} p_{ik}(u)q_{kj} = \infty \quad \text{for all } u \geq t.$$

The latter would contradict (2.7) since the left-hand side is finite for all t, so it is the former which holds. We know from the backward equation that $p_{ij}(t)$ is continuous for all i, j; hence by uniform convergence the integrand in (2.7) is continuous and we may differentiate to obtain

$$p'_{ij}(t) + p_{ij}(t)q_j = \sum_{k\neq j} p_{ik}(t)q_{kj}.$$

Hence $P(t)$ solves the forward equation.

To establish minimality let us suppose that $\widetilde{p}_{ij}(t)$ is another solution of the forward equation; then we also have

$$\widetilde{p}_{ij}(t) = \delta_{ij}e^{-q_i t} + \sum_{k\neq j}\int_0^t \widetilde{p}_{ik}(t-s) q_{kj} e^{-q_j s} ds.$$

A small variation of the argument leading to (2.6) shows that, for $n \geq 0$

$$\mathbb{P}_i(X_t = j, t < J_{n+1})$$
$$= \delta_{ij} e^{-q_i t} + \sum_{k \neq j} \int_0^t \mathbb{P}_i(X_t = j, t < J_n) q_{kj} e^{-q_j s} ds. \qquad (2.8)$$

If $\widetilde{P}(t) \geq 0$, then

$$\mathbb{P}(X_t = j, t < J_0) = 0 \leq \widetilde{p}_{ij}(t) \quad \text{for all } i, j \text{ and } t.$$

Let us suppose inductively that

$$\mathbb{P}_i(X_t = j, t < J_n) \leq \widetilde{p}_{ij}(t) \quad \text{for all } i, j \text{ and } t;$$

then by comparing (2.7) and (2.8) we obtain

$$\mathbb{P}_i(X_t = j, t < J_{n+1}) \leq \widetilde{p}_{ij}(t) \quad \text{for all } i, j \text{ and } t$$

and the induction proceeds. Hence

$$p_{ij}(t) = \lim_{n \to \infty} \mathbb{P}_i(X_t = j, t < J_n) \leq \widetilde{p}_{ij}(t) \quad \text{for all } i, j \text{ and } t. \qquad \square$$

Exercises

2.8.1 Two fleas are bound together to take part in a nine-legged race on the vertices A, B, C of a triangle. Flea 1 hops at random times in the clockwise direction; each hop takes the pair from one vertex to the next and the times between successive hops of Flea 1 are independent random variables, each with with exponential distribution, mean $1/\lambda$. Flea 2 behaves similarly, but hops in the anticlockwise direction, the times between his hops having mean $1/\mu$. Show that the probability that they are at A at a given time $t > 0$ (starting from A at time $t = 0$) is

$$\frac{1}{3} + \frac{2}{3} \exp\left\{-\frac{3(\lambda + \mu)t}{2}\right\} \cos\left\{\frac{\sqrt{3}(\lambda - \mu)t}{2}\right\}.$$

2.8.2 Let $(X_t)_{t \geq 0}$ be a birth-and-death process with rates $\lambda_n = n\lambda$ and $\mu_n = n\mu$, and assume that $X_0 = 1$. Show that $h(t) = \mathbb{P}(X_t = 0)$ satisfies

$$h(t) = \int_0^t e^{-(\lambda + \mu)s} \{\mu + \lambda h(t-s)^2\} ds$$

and deduce that if $\lambda \neq \mu$ then
$$h(t) = (\mu e^{\mu t} - \mu e^{\lambda t})/(\mu e^{\mu t} - \lambda e^{\lambda t}).$$

2.9 Non-minimal chains

This book concentrates entirely on processes which are right-continuous and minimal. These are the simplest sorts of process and, overwhelmingly, the ones of greatest practical application. We have seen in this chapter that we can associate to each distribution λ and Q-matrix Q a unique such process, the Markov chain with initial distribution λ and generator matrix Q. Indeed we have taken the liberty of defining Markov chains to be those processes which arise in this way. However, these processes do not by any means exhaust the class of memoryless continuous-time processes with values in a countable set I. There are many more exotic possibilities, the general theory of which goes very much deeper than the account given in this book. It is in the nature of things that these exotic cases have received the greater attention among mathematicians. Here are some examples to help you imagine the possibilities.

Example 2.9.1

Consider a birth process $(X_t)_{t \geq 0}$ starting from 0 with rates $q_i = 2^i$ for $i \geq 0$. We have chosen these rates so that
$$\sum_{i=0}^{\infty} q_i^{-1} = \sum_{i=0}^{\infty} 2^{-i} < \infty$$

which shows that the process explodes (see Theorems 2.3.2 and 2.5.2). We have until now insisted that $X_t = \infty$ for all $t \geq \zeta$, where ζ is the explosion time. But another obvious possibility is to start the process off again from 0 at time ζ, and do the same for all subsequent explosions. An argument based on the memoryless property of the exponential distribution shows that for $0 \leq t_0 \leq \ldots \leq t_{n+1}$ this process satisfies
$$\mathbb{P}(X_{t_{n+1}} = i_{n+1} \mid X_{t_0} = i_0, \ldots, X_{t_n} = i_n) = p_{i_n i_{n+1}}(t_{n+1} - t_n)$$

for a semigroup of stochastic matrices $(P(t) : t \geq 0)$ on I. This is the defining property for a more general class of Markov chains. Note that the chain is no longer determined by λ and Q alone; the rule for bringing $(X_t)_{t \geq 0}$ back into I after explosion also has to be given.

Example 2.9.2

We make a variation on the preceding example. Suppose now that the jump chain $(Y_n)_{n\geq 0}$ of $(X_t)_{t\geq 0}$ is the Markov chain on $\mathbb{Z}$ which moves one step away from 0 with probability $2/3$ and one step towards 0 with probability $1/3$, with $\pi_{01} = \pi_{0,-1} = 1/2$, and that $Y_0 = 0$. Let the transition rates for $(X_t)_{t\geq 0}$ be $q_i = 2^{|i|}$ for $i \in \mathbb{Z}$. Then $(X_t)_{t\geq 0}$ is again explosive. (A simple way to see this using some results of Chapter 3 is to check that $(Y_n)_{n\geq 0}$ is transient but $(X_t)_{t\geq 0}$ has an invariant distribution – by solution of the detailed balance equations. Then Theorem 3.5.3 makes explosion inevitable.) Now there are two ways in which $(X_t)_{t\geq 0}$ can explode, either $X_t \to -\infty$ or $X_t \to \infty$.

The process may again be restarted at 0 after explosion. Alternatively, we may choose the restart randomly, and according to the way that explosion occurred. For example

$$X_\zeta = \begin{cases} 0 & \text{if } X_t \to -\infty \text{ as } t \uparrow \zeta \\ Z & \text{if } X_t \to \infty \text{ as } t \uparrow \zeta \end{cases}$$

where Z takes values ± 1 with probability $1/2$.

Example 2.9.3

The processes in the preceding two examples, though no longer minimal, were at least right-continuous. Here is an altogether more exotic example, due to P. Lévy, which is not even right-continuous. Consider

$$D_n = \{k 2^{-n} : k \in \mathbb{Z}^+\} \qquad \text{for } n \geq 0$$

and set $I = \cup_n D_n$. With each $i \in D_n \backslash D_{n-1}$ we associate an independent exponential random variable S_i of parameter $(2^n)^2$. There are 2^{n-1} states in $(D^n \backslash D^{n-1}) \cap [0,1)$, so, for all $i \in I$

$$\mathbb{E}\left(\sum_{j \leq i} S_j\right) \leq (i+1) \sum_{n=0}^{\infty} 2^{n-1}(2^{-2n}) < \infty$$

and

$$\mathbb{P}\left(\sum_{j \leq i} S_j \to \infty \text{ as } i \to \infty\right) = 1.$$

Now define

$$X_t = \begin{cases} i & \text{if } \sum_{j<i} S_j \leq t < \sum_{j \leq i} S_j \text{ for some } i \in I \\ \infty & \text{otherwise.} \end{cases}$$

This process runs through all the dyadic rationals $i \in I$ in the usual order. It remains in $i \in D_n \backslash D_{n-1}$ for an exponential time of parameter 1. Between any two distinct states i and j it makes infinitely many visits to ∞. The Lebesgue measure of the set of times t when $X_t = \infty$ is zero. There is a semigroup of stochastic matrices $(P(t) : t \geq 0)$ on I such that, for $0 \leq t_0 \leq \ldots \leq t_{n+1}$

$$\mathbb{P}(X_{t_{n+1}} = i_{n+1} \mid X_{t_0} = i_0, \ldots, X_{t_n} = i_n) = p_{i_n i_{n+1}}(t_{n+1} - t_n).$$

In particular, $\mathbb{P}(X_t = \infty) = 0$ for all $t \geq 0$. The details may be found in *Markov Chains* by D. Freedman (Holden-Day, San Francisco, 1971).

We hope these three examples will serve to suggest some of the possibilities for more general continuous-time Markov chains. For further reading, see Freedman's book, or else *Markov Chains with Stationary Transition Probabilities* by K.-L. Chung (Springer, Berlin, 2nd edition, 1967), or *Diffusions, Markov Processes and Martingales, Vol 1: Foundations* by L. C. G. Rogers and D. Williams (Wiley, Chichester, 2nd edition, 1994).

2.10 Appendix: matrix exponentials

Define two norms on the space of real-valued $N \times N$-matrices

$$|Q| = \sup_{v \neq 0} |Qv|/|v|, \quad \|Q\|_\infty = \sup_{i,j} |q_{ij}|.$$

Obviously, $\|Q\|_\infty$ is finite for all Q and controls the size of the entries in Q. We shall show that the two norms are equivalent and that $|Q|$ is well adapted to sums and products of matrices, which $\|Q\|_\infty$ is not.

Lemma 2.10.1. *We have*

(a) $\|Q\|_\infty \leq |Q| \leq N\|Q\|_\infty$;

(b) $|Q_1 + Q_2| \leq |Q_1| + |Q_2|$ and $|Q_1 Q_2| \leq |Q_1||Q_2|$.

Proof. (a) For any vector v we have $|Qv| \leq |Q||v|$. In particular, for the vector $\varepsilon_j = (0, \ldots, 1, \ldots, 0)$, with 1 in the jth place, we have $|Q\varepsilon_j| \leq |Q|$. The supremum defining $\|Q\|_\infty$ is achieved, at j say, so

$$\|Q\|_\infty^2 \leq \sum_i (q_{ij})^2 = |Q\varepsilon_j|^2 \leq |Q|^2.$$

On the other hand

$$|Qv|^2 = \sum_i \left(\sum_j q_{ij} v_j\right)^2$$
$$\leq \sum_i \left(\sum_j \|Q\|_\infty |v_j|\right)^2$$
$$= N\|Q\|_\infty^2 \left(\sum_j |v_j|\right)^2$$

and, by the Cauchy–Schwarz inequality

$$\left(\sum_j |v_j|\right)^2 \leq N \sum_j v_j^2$$

so $|Qv|^2 \leq N^2 \|Q\|_\infty^2 |v|^2$. This implies that $|Q| \leq N\|Q\|_\infty$.

(b) For any vector v we have

$$|(Q_1 + Q_2)v| \leq |Q_1 v| + |Q_2 v| \leq (|Q_1| + |Q_2|)|v|,$$
$$|Q_1 Q_2 v| \leq |Q_1||Q_2 v| \leq |Q_1||Q_2||v|. \qquad \square$$

Now for $n = 0, 1, 2, \ldots$, consider the finite sum

$$E(n) = \sum_{k=0}^n \frac{Q^k}{k!}.$$

For each i and j, and $m \leq n$, we have

$$|(E(n) - E(m))_{ij}| \leq \|E(n) - E(m)\|_\infty \leq |E(n) - E(m)|$$
$$= \left|\sum_{k=m+1}^n \frac{Q^k}{k!}\right|$$
$$\leq \sum_{k=m+1}^n \frac{|Q|^k}{k!}.$$

Since $|Q| \leq N\|Q\|_\infty < \infty$, $\sum_{k=0}^\infty |Q|^k/k!$ converges by the ratio test, so

$$\sum_{k=m+1}^n \frac{|Q|^k}{k!} \longrightarrow 0 \quad \text{as } m, n \to \infty.$$

2.10 Appendix: matrix exponentials

Hence each component of $E(n)$ forms a Cauchy sequence, which therefore converges, proving that

$$e^Q = \sum_{k=0}^{\infty} \frac{Q^k}{k!}$$

is well defined and, indeed, that the power series

$$(e^{tQ})_{ij} = \sum_{k=0}^{\infty} \frac{(tQ)^k_{ij}}{k!}$$

has infinite radius of convergence for all i, j. Finally, for two commuting matrices Q_1 and Q_2 we have

$$\begin{aligned}
e^{Q_1+Q_2} &= \sum_{n=0}^{\infty} \frac{(Q_1+Q_2)^n}{n!} \\
&= \sum_{n=0}^{\infty} \frac{1}{n!} \sum_{k=0}^{n} \frac{n!}{k!(n-k)!} Q_1^k Q_2^{n-k} \\
&= \sum_{k=0}^{\infty} \frac{Q_1^k}{k!} \sum_{n=k}^{\infty} \frac{Q_2^{n-k}}{(n-k)!} \\
&= e^{Q_1} e^{Q_2}.
\end{aligned}$$

3
Continuous-time Markov chains II

This chapter brings together the discrete-time and continuous-time theories, allowing us to deduce analogues, for continuous-time chains, of all the results given in Chapter 1. All the facts from Chapter 2 that are necessary to understand this synthesis are reviewed in Section 3.1. You will require a reasonable understanding of Chapter 1 here, but, given such an understanding, this chapter should look reassuringly familiar. Exercises remain an important part of the text.

3.1 Basic properties

Let I be a countable set. Recall that a *Q-matrix* on I is a matrix $Q = (q_{ij} : i, j \in I)$ satisfying the following conditions:

(i) $0 \leq -q_{ii} < \infty$ for all i;
(ii) $q_{ij} \geq 0$ for all $i \neq j$;
(iii) $\sum_{j \in I} q_{ij} = 0$ for all i.

We set $q_i = q(i) = -q_{ii}$. Associated to any Q-matrix is a *jump matrix* $\Pi = (\pi_{ij} : i, j \in I)$ given by

$$\pi_{ij} = \begin{cases} q_{ij}/q_i & \text{if } j \neq i \text{ and } q_i \neq 0 \\ 0 & \text{if } j \neq i \text{ and } q_i = 0, \end{cases}$$

$$\pi_{ii} = \begin{cases} 0 & \text{if } q_i \neq 0 \\ 1 & \text{if } q_i = 0. \end{cases}$$

Note that Π is a stochastic matrix.

3.1 Basic properties

A *sub-stochastic matrix* on I is a matrix $P = (p_{ij} : i, j \in I)$ with non-negative entries and such that

$$\sum_{j \in I} p_{ij} \leq 1 \quad \text{for all} \quad i.$$

Associated to any Q-matrix is a *semigroup* $(P(t) : t \geq 0)$ of sub-stochastic matrices $P(t) = (p_{ij}(t) : i, j \in I)$. As the name implies we have

$$P(s)P(t) = P(s+t) \quad \text{for all} \quad s, t \geq 0.$$

You will need to be familiar with the following terms introduced in Section 2.2: *minimal right-continuous random process, jump times, holding times, jump chain* and *explosion*. Briefly, a right-continuous process $(X_t)_{t \geq 0}$ runs through a sequence of states $Y_0, Y_1, Y_2, \ldots$, being held in these states for times $S_1, S_2, S_3, \ldots$ respectively and jumping to the next state at times $J_1, J_2, J_3, \ldots$. Thus $J_n = S_1 + \ldots + S_n$. The discrete-time process $(Y_n)_{n \geq 0}$ is the jump chain, $(S_n)_{n \geq 1}$ are the holding times and $(J_n)_{n \geq 1}$ are the jump times. The explosion time ζ is given by

$$\zeta = \sum_{n=1}^{\infty} S_n = \lim_{n \to \infty} J_n.$$

For a minimal process we take a new state ∞ and insist that $X_t = \infty$ for all $t \geq \zeta$. An important point is that a minimal right-continuous process is determined by its jump chain and holding times.

The data for a continuous-time Markov chain $(X_t)_{t \geq 0}$ are a distribution λ and a Q-matrix Q. The distribution λ gives the *initial distribution*, the distribution of X_0. The Q-matrix is known as the *generator matrix* of $(X_t)_{t \geq 0}$ and determines how the process evolves from its initial state. We established in Section 2.8 that there are two different, but equivalent, ways to describe how the process evolves.

The first, in terms of jump chain and holding times, states that
(a) $(Y_n)_{n \geq 0}$ is Markov(λ, Π);
(b) conditional on $Y_0 = i_0, \ldots, Y_{n-1} = i_{n-1}$, the holding times $S_1, \ldots, S_n$ are independent exponential random variables of parameters $q_{i_0}, \ldots, q_{i_{n-1}}$.

Put more simply, given that the chain starts at i, it waits there for an exponential time of parameter q_i and then jumps to a new state, choosing state j with probability π_{ij}. It then starts afresh, forgetting what has gone before.

The second description, in terms of the semigroup, states that the finite-dimensional distributions of the process are given by

(c) for all $n = 0, 1, 2, \ldots$, all times $0 \leq t_0 \leq t_1 \leq \cdots \leq t_{n+1}$ and all states $i_0, i_1, \ldots, i_{n+1}$

$$\mathbb{P}(X_{t_{n+1}} = i_{n+1} \mid X_{t_0} = i_0, \ldots, X_{t_n} = i_n) = p_{i_n i_{n+1}}(t_{n+1} - t_n).$$

Again, put more simply, given that the chain starts at i, by time t it is found in state j with probability $p_{ij}(t)$. It then starts afresh, forgetting what has gone before. In the case where

$$\widetilde{p}_{i\infty}(t) := 1 - \sum_{j \in I} p_{ij}(t) > 0$$

the chain is found at ∞ with probability $\widetilde{p}_{i\infty}(t)$. The semigroup $P(t)$ is referred to as the *transition matrix* of the chain and its entries $p_{ij}(t)$ are the *transition probabilities*. This description implies that for all $h > 0$ the discrete skeleton $(X_{nh})_{n \geq 0}$ is Markov$(\lambda, P(h))$. Strictly, in the explosive case, that is, when $P(t)$ is strictly sub-stochastic, we should say Markov$(\widetilde{\lambda}, \widetilde{P}(h))$, where $\widetilde{\lambda}$ and $\widetilde{P}(h)$ are defined on $I \cup \{\infty\}$, extending λ and $P(h)$ by $\widetilde{\lambda}_\infty = 0$ and $\widetilde{p}_{\infty j}(h) = 0$. But there is no danger of confusion in using the simpler notation.

The information coming from these two descriptions is sufficient for most of the analysis of continuous-time chains done in this chapter. Note that we have not yet said how the semigroup $P(t)$ is associated to the Q-matrix Q, except via the process! This extra information will be required when we discuss reversibility in Section 3.7. So we recall from Section 2.8 that the semigroup is characterized as the minimal non-negative solution of the *backward equation*

$$P'(t) = QP(t), \quad P(0) = I$$

which reads in components

$$p'_{ij}(t) = \sum_{k \in I} q_{ik} p_{kj}(t), \quad p_{ij}(0) = \delta_{ij}.$$

The semigroup is also the minimal non-negative solution of the *forward equation*

$$P'(t) = P(t)Q, \quad P(0) = I.$$

In the case where I is finite, $P(t)$ is simply the matrix exponential e^{tQ}, and is the *unique* solution of the backward and forward equations.

3.2 Class structure

A first step in the analysis of a continuous-time Markov chain $(X_t)_{t\geq 0}$ is to identify its class structure. We emphasise that we deal only with minimal chains, those that die after explosion. Then the class structure is simply the discrete-time class structure of the jump chain $(Y_n)_{n\geq 0}$, as discussed in Section 1.2.

We say that *i leads to j* and write $i \to j$ if
$$\mathbb{P}_i(X_t = j \text{ for some } t \geq 0) > 0.$$

We say *i communicates with j* and write $i \leftrightarrow j$ if both $i \to j$ and $j \to i$. The notions of *communicating class*, *closed class*, *absorbing state* and *irreducibility* are inherited from the jump chain.

Theorem 3.2.1. *For distinct states i and j the following are equivalent:*
  (i) $i \to j$;
  (ii) $i \to j$ for the jump chain;
  (iii) $q_{i_0 i_1} q_{i_1 i_2} \cdots q_{i_{n-1} i_n} > 0$ for some states $i_0, i_1, \ldots, i_n$ with $i_0 = i$, $i_n = j$;
  (iv) $p_{ij}(t) > 0$ for all $t > 0$;
  (v) $p_{ij}(t) > 0$ for some $t > 0$.

Proof. Implications (iv) $\Rightarrow$ (v) $\Rightarrow$ (i) $\Rightarrow$ (ii) are clear. If (ii) holds, then by Theorem 1.2.1, there are states $i_0, i_1, \ldots, i_n$ with $i_0 = i$, $i_n = j$ and $\pi_{i_0 i_1} \pi_{i_1 i_2} \cdots \pi_{i_{n-1} i_n} > 0$, which implies (iii). If $q_{ij} > 0$, then
$$p_{ij}(t) \geq \mathbb{P}_i(J_1 \leq t, Y_1 = j, S_2 > t) = (1 - e^{-q_i t})\pi_{ij} e^{-q_j t} > 0$$
for all $t > 0$, so if (iii) holds, then
$$p_{ij}(t) \geq p_{i_0 i_1}(t/n) \cdots p_{i_{n-1} i_n}(t/n) > 0$$
for all $t > 0$, and (iv) holds. $\square$

Condition (iv) of Theorem 3.2.1 shows that the situation is simpler than in discrete-time, where it may be possible to reach a state, but only after a certain length of time, and then only periodically.

3.3 Hitting times and absorption probabilities

Let $(X_t)_{t\geq 0}$ be a Markov chain with generator matrix Q. The *hitting time* of a subset A of I is the random variable D^A defined by
$$D^A(\omega) = \inf\{t \geq 0 : X_t(\omega) \in A\}$$

with the usual convention that $\inf \emptyset = \infty$. We emphasise that $(X_t)_{t \geq 0}$ is minimal. So if H^A is the hitting time of A for the jump chain, then

$$\{H^A < \infty\} = \{D^A < \infty\}$$

and on this set we have

$$D^A = J_{H^A}.$$

The probability, starting from i, that $(X_t)_{t \geq 0}$ ever hits A is then

$$h_i^A = \mathbb{P}_i(D^A < \infty) = \mathbb{P}_i(H^A < \infty).$$

When A is a closed class, h_i^A is called the *absorption probability*. Since the hitting probabilities are those of the jump chain we can calculate them as in Section 1.3.

Theorem 3.3.1. *The vector of hitting probabilities $h^A = (h_i^A : i \in I)$ is the minimal non-negative solution to the system of linear equations*

$$\begin{cases} h_i^A = 1 & \text{for } i \in A, \\ \sum_{j \in I} q_{ij} h_j^A = 0 & \text{for } i \notin A. \end{cases}$$

Proof. Apply Theorem 1.3.2 to the jump chain and rewrite (1.3) in terms of Q. □

The average time taken, starting from i, for $(X_t)_{t \geq 0}$ to reach A is given by

$$k_i^A = \mathbb{E}_i(D^A).$$

In calculating k_i^A we have to take account of the holding times so the relationship to the discrete-time case is not quite as simple.

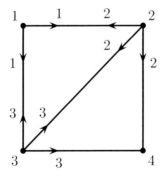

3.3 Hitting times and absorption probabilities

Example 3.3.2

Consider the Markov chain $(X_t)_{t\geq 0}$ with the diagram given on the preceding page. How long on average does it take to get from 1 to 4?

Set $k_i = \mathbb{E}_i(\text{time to get to 4})$. On starting in 1 we spend an average time $q_1^{-1} = 1/2$ in 1, then jump with equal probability to 2 or 3. Thus

$$k_1 = \tfrac{1}{2} + \tfrac{1}{2}k_2 + \tfrac{1}{2}k_3$$

and similarly

$$k_2 = \tfrac{1}{6} + \tfrac{1}{3}k_1 + \tfrac{1}{3}k_3, \quad k_3 = \tfrac{1}{9} + \tfrac{1}{3}k_1 + \tfrac{1}{3}k_2.$$

On solving these linear equations we find $k_1 = 17/12$.

Here is the general result. The proof follows the same lines as Theorem 1.3.5.

Theorem 3.3.3. *Assume that $q_i > 0$ for all $i \notin A$. The vector of expected hitting times $k^A = (k_i^A : i \in I)$ is the minimal non-negative solution to the system of linear equations*

$$\begin{cases} k_i^A = 0 & \text{for } i \in A \\ -\sum_{j \in I} q_{ij} k_j^A = 1 & \text{for } i \notin A. \end{cases} \quad (3.1)$$

Proof. First we show that k^A satisfies (3.1). If $X_0 = i \in A$, then $D^A = 0$, so $k_i^A = 0$. If $X_0 = i \notin A$, then $D^A \geq J_1$, so by the Markov property of the jump chain

$$\mathbb{E}_i(D^A - J_1 \mid Y_1 = j) = \mathbb{E}_j(D^A),$$

so

$$k_i^A = \mathbb{E}_i(D^A) = \mathbb{E}_i(J_1) + \sum_{j \neq i} \mathbb{E}(D^A - J_1 \mid Y_1 = j)\mathbb{P}_i(Y_1 = j) = q_i^{-1} + \sum_{j \neq i} \pi_{ij} k_j^A$$

and so

$$-\sum_{j \in I} q_{ij} k_j^A = 1.$$

Suppose now that $y = (y_i : i \in I)$ is another solution to (3.1). Then $k_i^A = y_i = 0$ for $i \in A$. Suppose $i \notin A$, then

$$y_i = q_i^{-1} + \sum_{j \notin A} \pi_{ij} y_j = q_i^{-1} + \sum_{j \notin A} \pi_{ij} \left(q_j^{-1} + \sum_{k \notin A} \pi_{jk} y_k \right)$$

$$= \mathbb{E}_i(S_1) + \mathbb{E}_i(S_2 1_{\{H^A \geq 2\}}) + \sum_{j \notin A} \sum_{k \notin A} \pi_{ij} \pi_{jk} y_k.$$

By repeated substitution for y in the final term we obtain after n steps

$$y_i = \mathbb{E}_i(S_1) + \cdots + \mathbb{E}_i(S_n 1_{\{H^A \geq n\}}) + \sum_{j_1 \notin A} \cdots \sum_{j_n \notin A} \pi_{ij_1} \cdots \pi_{j_{n-1} j_n} y_{j_n}.$$

So, if y is non-negative

$$y_i \geq \sum_{m=1}^n \mathbb{E}_i(S_m 1_{H^A \geq m}) = \mathbb{E}_i\left(\sum_{m=1}^{H^A \wedge n} S_m\right)$$

where we use the notation $H^A \wedge n$ for the minimum of H^A and n. Now

$$\sum_{m=1}^{H^A} S_m = D_A$$

so, by monotone convergence, $y_i \geq \mathbb{E}_i(D_A) = k_i^A$, as required. $\square$

Exercise

3.3.1 Consider the Markov chain on $\{1, 2, 3, 4\}$ with generator matrix

$$Q = \begin{pmatrix} -1 & 1/2 & 1/2 & 0 \\ 1/4 & -1/2 & 0 & 1/4 \\ 1/6 & 0 & -1/3 & 1/6 \\ 0 & 0 & 0 & 0 \end{pmatrix}.$$

Calculate (a) the probability of hitting 3 starting from 1, (b) the expected time to hit 4 starting from 1.

3.4 Recurrence and transience

Let $(X_t)_{t \geq 0}$ be Markov chain with generator matrix Q. Recall that we insist $(X_t)_{t \geq 0}$ be minimal. We say a state i is *recurrent* if

$$\mathbb{P}_i(\{t \geq 0 : X_t = i\} \text{ is unbounded}) = 1.$$

We say that i is *transient* if

$$\mathbb{P}_i(\{t \geq 0 : X_t = i\} \text{ is unbounded}) = 0.$$

Note that if $(X_t)_{t \geq 0}$ can explode starting from i then i is certainly not recurrent. The next result shows that, like class structure, recurrence and transience are determined by the jump chain.

Theorem 3.4.1. *We have:*
  (i) *if i is recurrent for the jump chain $(Y_n)_{n \geq 0}$, then i is recurrent for $(X_t)_{t \geq 0}$;*
  (ii) *if i is transient for the jump chain, then i is transient for $(X_t)_{t \geq 0}$;*
  (iii) *every state is either recurrent or transient;*
  (iv) *recurrence and transience are class properties.*

Proof. (i) Suppose i is recurrent for $(Y_n)_{n \geq 0}$. If $X_0 = i$ then $(X_t)_{t \geq 0}$ does not explode and $J_n \to \infty$ by Theorem 2.7.1. Also $X(J_n) = Y_n = i$ infinitely often, so $\{t \geq 0 : X_t = i\}$ is unbounded, with probability 1.
(ii) Suppose i is transient for $(Y_n)_{n \geq 0}$. If $X_0 = i$ then
$$N = \sup\{n \geq 0 : Y_n = i\} < \infty,$$
so $\{t \geq 0 : X_t = i\}$ is bounded by $J(N+1)$, which is finite, with probability 1, because $(Y_n : n \leq N)$ cannot include an absorbing state.
(iii) Apply Theorem 1.5.3 to the jump chain.
(iv) Apply Theorem 1.5.4 to the jump chain. □

The next result gives continuous-time analogues of the conditions for recurrence and transience found in Theorem 1.5.3. We denote by T_i the *first passage time* of $(X_t)_{t \geq 0}$ to state i, defined by
$$T_i(\omega) = \inf\{t \geq J_1(\omega) : X_t(\omega) = i\}.$$

Theorem 3.4.2. *The following dichotomy holds:*
  (i) *if $q_i = 0$ or $\mathbb{P}_i(T_i < \infty) = 1$, then i is recurrent and $\int_0^\infty p_{ii}(t)dt = \infty$;*
  (ii) *if $q_i > 0$ and $\mathbb{P}_i(T_i < \infty) < 1$, then i is transient and $\int_0^\infty p_{ii}(t)dt < \infty$.*

Proof. If $q_i = 0$, then $(X_t)_{t \geq 0}$ cannot leave i, so i is recurrent, $p_{ii}(t) = 1$ for all t, and $\int_0^\infty p_{ii}(t)dt = \infty$. Suppose then that $q_i > 0$. Let N_i denote the first passage time of the jump chain $(Y_n)_{n \geq 0}$ to state i. Then
$$\mathbb{P}_i(N_i < \infty) = \mathbb{P}_i(T_i < \infty)$$
so i is recurrent if and only if $\mathbb{P}_i(T_i < \infty) = 1$, by Theorem 3.4.1 and the corresponding result for the jump chain.

Write $\pi_{ij}^{(n)}$ for the (i,j) entry in Π^n. We shall show that
$$\int_0^\infty p_{ii}(t)dt = \frac{1}{q_i} \sum_{n=0}^\infty \pi_{ii}^{(n)} \qquad (3.2)$$
so that i is recurrent if and only if $\int_0^\infty p_{ii}(t)dt = \infty$, by Theorem 3.4.1 and the corresponding result for the jump chain.

To establish (3.2) we use Fubini's theorem (see Section 6.4):

$$\int_0^\infty p_{ii}(t)dt = \int_0^\infty \mathbb{E}_i(1_{\{X_t=i\}})dt = \mathbb{E}_i \int_0^\infty 1_{\{X_t=i\}}dt$$

$$= \mathbb{E}_i \sum_{n=0}^\infty S_{n+1} 1_{\{Y_n=i\}}$$

$$= \sum_{n=0}^\infty \mathbb{E}_i(S_{n+1} \mid Y_n = i)\mathbb{P}_i(Y_n = i) = \frac{1}{q_i}\sum_{n=0}^\infty \pi_{ii}^{(n)}. \qquad \square$$

Finally, we show that recurrence and transience are determined by any discrete-time sampling of $(X_t)_{t\geq 0}$.

Theorem 3.4.3. Let $h > 0$ be given and set $Z_n = X_{nh}$.
(i) If i is recurrent for $(X_t)_{t\geq 0}$ then i is recurrent for $(Z_n)_{n\geq 0}$.
(ii) If i is transient for $(X_t)_{t\geq 0}$ then i is transient for $(Z_n)_{n\geq 0}$.

Proof. Claim (ii) is obvious. To prove (i) we use for $nh \leq t < (n+1)h$ the estimate

$$p_{ii}((n+1)h) \geq e^{-q_i h} p_{ii}(t)$$

which follows from the Markov property. Then, by monotone convergence

$$\int_0^\infty p_{ii}(t)dt \leq h e^{q_i h} \sum_{n=1}^\infty p_{ii}(nh)$$

and the result follows by Theorems 1.5.3 and 3.4.2. $\square$

Exercise

3.4.1 Customers arrive at a certain queue in a Poisson process of rate λ. The single 'server' has two states A and B, state A signifying that he is 'in attendance' and state B that he is having a tea-break. Independently of how many customers are in the queue, he fluctuates between these states as a Markov chain Y on $\{A, B\}$ with Q-matrix

$$\begin{pmatrix} -\alpha & \alpha \\ \beta & -\beta \end{pmatrix}.$$

The total service time for any customer is exponentially distributed with parameter μ and is independent of the chain Y and of the service times of other customers.

Describe the system as a Markov chain X with state-space

$$\{A_0, A_1, A_2, \ldots\} \cup \{B_0, B_1, B_2, \ldots\},$$

A_n signifying that the server is in state A and there are n people in the queue (including anyone being served) and B_n signifying that the server is in state B and there are n people in the queue.

Explain why, for some θ in $(0, 1]$, and $k = 0, 1, 2, \ldots$,

$$P(X \text{ hits } A_0 | X_0 = A_k) = \theta^k.$$

Show that $(\theta - 1)f(\theta) = 0$, where

$$f(\theta) = \lambda^2 \theta^2 - \lambda(\lambda + \mu + \alpha + \beta)\theta + (\lambda + \beta)\mu.$$

By considering $f(1)$ or otherwise, prove that X is transient if $\mu\beta < \lambda(\alpha+\beta)$, and explain why this is intuitively obvious.

3.5 Invariant distributions

Just as in the discrete-time theory, the notions of invariant distribution and measure play an important role in the study of continuous-time Markov chains. We say that λ is *invariant* if

$$\lambda Q = 0.$$

Theorem 3.5.1. *Let Q be a Q-matrix with jump matrix Π and let λ be a measure. The following are equivalent:*

(i) λ *is invariant;*
(ii) $\mu\Pi = \mu$ *where* $\mu_i = \lambda_i q_i$.

Proof. We have $q_i(\pi_{ij} - \delta_{ij}) = q_{ij}$ for all i, j, so

$$(\mu(\Pi - I))_j = \sum_{i \in I} \mu_i(\pi_{ij} - \delta_{ij}) = \sum_{i \in I} \lambda_i q_{ij} = (\lambda Q)_j. \qquad \square$$

This tie-up with measures invariant for the jump matrix means that we can use the existence and uniqueness results of Section 1.7 to obtain the following result.

Theorem 3.5.2. *Suppose that Q is irreducible and recurrent. Then Q has an invariant measure λ which is unique up to scalar multiples.*

Proof. Let us exclude the trivial case $I = \{i\}$; then irreducibility forces $q_i > 0$ for all i. By Theorems 3.2.1 and 3.4.1, Π is irreducible and recurrent. Then, by Theorems 1.7.5 and 1.7.6, Π has an invariant measure μ, which is unique up to scalar multiples. So, by Theorem 3.5.1, we can take $\lambda_i = \mu_i/q_i$ to obtain an invariant measure unique up to scalar multiples. □

Recall that a state i is recurrent if $q_i = 0$ or $\mathbb{P}_i(T_i < \infty) = 1$. If $q_i = 0$ or the *expected return time* $m_i = \mathbb{E}_i(T_i)$ is finite then we say i is *positive recurrent*. Otherwise a recurrent state i is called *null recurrent*. As in the discrete-time case positive recurrence is tied up with the existence of an invariant distribution.

Theorem 3.5.3. *Let Q be an irreducible Q-matrix. Then the following are equivalent:*
  (i) *every state is positive recurrent;*
  (ii) *some state i is positive recurrent;*
  (iii) *Q is non-explosive and has an invariant distribution λ.*

Moreover, when (iii) holds we have $m_i = 1/(\lambda_i q_i)$ for all i.

Proof. Let us exclude the trivial case $I = \{i\}$; then irreducibility forces $q_i > 0$ for all i. It is obvious that (i) implies (ii). Define $\mu^i = (\mu^i_j : j \in I)$ by
$$\mu^i_j = \mathbb{E}_i \int_0^{T_i \wedge \zeta} 1_{\{X_s = j\}} ds,$$
where $T_i \wedge \zeta$ denotes the minimum of T_i and ζ. By monotone convergence,
$$\sum_{j \in I} \mu^i_j = \mathbb{E}_i(T_i \wedge \zeta).$$
Denote by N_i the first passage time of the jump chain to state i. By Fubini's theorem
$$\mu^i_j = \mathbb{E}_i \sum_{n=0}^{\infty} S_{n+1} 1_{\{Y_n = j, n < N_i\}}$$
$$= \sum_{n=0}^{\infty} \mathbb{E}_i(S_{n+1} \mid Y_n = j) \mathbb{E}_i(1_{\{Y_n = j, n < N_i\}})$$
$$= q_j^{-1} \mathbb{E}_i \sum_{n=0}^{\infty} 1_{\{Y_n = j, n < N_i\}}$$
$$= q_j^{-1} \mathbb{E}_i \sum_{n=0}^{N_i - 1} 1_{\{Y_n = j\}} = \gamma^i_j/q_j$$

3.5 Invariant distributions

where, in the notation of Section 1.7, γ_j^i is the expected time in j between visits to i for the jump chain.

Suppose (ii) holds, then i is certainly recurrent, so the jump chain is recurrent, and Q is non-explosive, by Theorem 2.7.1. We know that $\gamma^i \Pi = \gamma^i$ by Theorem 1.7.5, so $\mu^i Q = 0$ by Theorem 3.5.1. But μ^i has finite total mass

$$\sum_{j \in I} \mu_j^i = \mathbb{E}_i(T_i) = m_i$$

so we obtain an invariant distribution λ by setting $\lambda_j = \mu_j^i/m_i$.

On the other hand, suppose (iii) holds. Fix $i \in I$ and set $\nu_j = \lambda_j q_j/(\lambda_i q_i)$; then $\nu_i = 1$ and $\nu \Pi = \nu$ by Theorem 3.5.1, so $\nu_j \geq \gamma_j^i$ for all j by Theorem 1.7.6. So

$$m_i = \sum_{j \in I} \mu_j^i = \sum_{j \in I} \gamma_j^i/q_j \leq \sum_{j \in I} \nu_j/q_j$$
$$= \sum_{j \in I} \lambda_j/(\lambda_i q_i) = 1/(\lambda_i q_i) < \infty$$

showing that i is positive recurrent.

To complete the proof we return to the preceding calculation armed with the knowledge that Q is recurrent, hence Π is recurrent, $\nu_j = \gamma_j^i$ and $m_i = 1/(\lambda_i q_i)$ for all i. □

The following example is a caution that the existence of an invariant distribution for a continuous-time Markov chain is not enough to guarantee positive recurrence, or even recurrence.

Example 3.5.4

Consider the Markov chain $(X_t)_{t \geq 0}$ on $\mathbb{Z}^+$ with the following diagram, where $q_i > 0$ for all i and where $0 < \lambda = 1 - \mu < 1$:

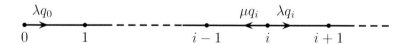

The jump chain behaves as a simple random walk away from 0, so $(X_t)_{t \geq 0}$ is recurrent if $\lambda \leq \mu$ and transient if $\lambda > \mu$. To compute an invariant measure ν it is convenient to use the *detailed balance equations*

$$\nu_i q_{ij} = \nu_j q_{ji} \qquad \text{for all } i, j.$$

Look ahead to Lemma 3.7.2 to see that any solution is invariant. In this case the non-zero equations read

$$\nu_i \lambda q_i = \nu_{i+1} \mu q_{i+1} \qquad \text{for all } i.$$

So a solution is given by $\nu_i = q_i^{-1}(\lambda/\mu)^i$. If the jump rates q_i are constant then ν can be normalized to produce an invariant distribution precisely when $\lambda < \mu$.

Consider, on the other hand, the case where $q_i = 2^i$ for all i and $1 < \lambda/\mu < 2$. Then ν has finite total mass so $(X_t)_{t \geq 0}$ has an invariant distribution, but $(X_t)_{t \geq 0}$ is also transient. Given Theorem 3.5.3, the only possibility is that $(X_t)_{t \geq 0}$ is explosive.

The next result justifies calling measures λ with $\lambda Q = 0$ invariant.

Theorem 3.5.5. *Let Q be irreducible and recurrent, and let λ be a measure. Let $s > 0$ be given. The following are equivalent:*

(i) $\lambda Q = 0$;
(ii) $\lambda P(s) = \lambda$.

Proof. There is a very simple proof in the case of finite state-space: by the backward equation

$$\frac{d}{ds}\lambda P(s) = \lambda P'(s) = \lambda Q P(s)$$

so $\lambda Q = 0$ implies $\lambda P(s) = \lambda P(0) = \lambda$ for all s; $P(s)$ is also recurrent, so $\mu P(s) = \mu$ implies that μ is proportional to λ, so $\mu Q = 0$.

For infinite state-space, the interchange of differentiation with the summation involved in multiplication by λ is not justified and an entirely different proof is needed.

Since Q is recurrent, it is non-explosive by Theorem 2.7.1, and $P(s)$ is recurrent by Theorem 3.4.3. Hence any λ satisfying (i) or (ii) is unique up to scalar multiples; and from the proof of Theorem 3.5.3, if we fix i and set

$$\mu_j = \mathbb{E}_i \int_0^{T_i} 1_{\{X_t = j\}} dt,$$

then $\mu Q = 0$. Thus it suffices to show $\mu P(s) = \mu$. By the strong Markov property at T_i (which is a simple consequence of the strong Markov property of the jump chain)

$$\mathbb{E}_i \int_0^s 1_{\{X_t = j\}} dt = \mathbb{E}_i \int_{T_i}^{T_i + s} 1_{\{X_t = j\}} dt.$$

Hence, using Fubini's theorem,

$$\mu_j = \mathbb{E}_i \int_s^{s+T_i} 1_{\{X_t = j\}} dt$$

$$= \int_0^\infty \mathbb{P}_i(X_{s+t} = j, t < T_i) dt$$

$$= \int_0^\infty \sum_{k \in I} \mathbb{P}_i(X_t = k, t < T_i) p_{kj}(s) dt$$

$$= \sum_{k \in I} \left(\mathbb{E}_i \int_0^{T_i} 1_{\{X_t = k\}} dt \right) p_{kj}(s)$$

$$= \sum_{k \in I} \mu_k p_{kj}(s)$$

as required. □

Theorem 3.5.6. *Let Q be an irreducible non-explosive Q-matrix having an invariant distribution λ. If $(X_t)_{t \geq 0}$ is Markov(λ, Q) then so is $(X_{s+t})_{t \geq 0}$ for any $s \geq 0$.*

Proof. By Theorem 3.5.5, for all i,

$$\mathbb{P}(X_s = i) = (\lambda P(s))_i = \lambda_i$$

so, by the Markov property, conditional on $X_s = i$, $(X_{s+t})_{t \geq 0}$ is Markov(δ_i, Q). □

3.6 Convergence to equilibrium

We now investigate the limiting behaviour of $p_{ij}(t)$ as $t \to \infty$ and its relation to invariant distributions. You will see that the situation is analogous to the case of discrete-time, only there is no longer any possibility of periodicity.

We shall need the following estimate of uniform continuity for the transition probabilities.

Lemma 3.6.1. *Let Q be a Q-matrix with semigroup $P(t)$. Then, for all $t, h \geq 0$*

$$|p_{ij}(t+h) - p_{ij}(t)| \leq 1 - e^{-q_i h}.$$

Proof. We have

$$|p_{ij}(t+h) - p_{ij}(t)| = \left| \sum_{k \in I} p_{ik}(h) p_{kj}(t) - p_{ij}(t) \right|$$

$$= \left| \sum_{k \neq i} p_{ik}(h) p_{kj}(t) - (1 - p_{ii}(h)) p_{ij}(t) \right|$$

$$\leq 1 - p_{ii}(h) \leq \mathbb{P}_i(J_1 \leq h) = 1 - e^{-q_i h}. \quad \square$$

Theorem 3.6.2 (Convergence to equilibrium). *Let Q be an irreducible non-explosive Q-matrix with semigroup $P(t)$, and having an invariant distribution λ. Then for all states i, j we have*

$$p_{ij}(t) \to \lambda_j \quad \text{as} \quad t \to \infty.$$

Proof. Let $(X_t)_{t \geq 0}$ be Markov(δ_i, Q). Fix $h > 0$ and consider the *h-skeleton* $Z_n = X_{nh}$. By Theorem 2.8.4

$$\mathbb{P}(Z_{n+1} = i_{n+1} \mid Z_0 = i_0, \ldots, Z_n = i_n) = p_{i_n i_{n+1}}(h)$$

so $(Z_n)_{n \geq 0}$ is discrete-time Markov$(\delta_i, P(h))$. By Theorem 3.2.1 irreducibility implies $p_{ij}(h) > 0$ for all i, j so $P(h)$ is irreducible and aperiodic. By Theorem 3.5.5, λ is invariant for $P(h)$. So, by discrete-time convergence to equilibrium, for all i, j

$$p_{ij}(nh) \to \lambda_j \quad \text{as} \quad n \to \infty.$$

Thus we have a lattice of points along which the desired limit holds; we fill in the gaps using uniform continuity. Fix a state i. Given $\varepsilon > 0$ we can find $h > 0$ so that

$$1 - e^{-q_i s} \leq \varepsilon/2 \quad \text{for } 0 \leq s \leq h$$

and then find N, so that

$$|p_{ij}(nh) - \lambda_j| \leq \varepsilon/2 \quad \text{for } n \geq N.$$

For $t \geq Nh$ we have $nh \leq t < (n+1)h$ for some $n \geq N$ and

$$|p_{ij}(t) - \lambda_j| \leq |p_{ij}(t) - p_{ij}(nh)| + |p_{ij}(nh) - \lambda_j| \leq \varepsilon$$

by Lemma 3.6.1. Hence

$$p_{ij}(t) \to \lambda_j \quad \text{as} \quad n \to \infty. \qquad \square$$

The complete description of limiting behaviour for irreducible chains in continuous time is provided by the following result. It follows from Theorem 1.8.5 by the same argument we used in the preceding result. We do not give the details.

Theorem 3.6.3. Let Q be an irreducible Q-matrix and let ν be any distribution. Suppose that $(X_t)_{t\geq 0}$ is Markov(ν, Q). Then

$$\mathbb{P}(X_t = j) \to 1/(q_j m_j) \quad \text{as } t \to \infty \quad \text{for all } j \in I$$

where m_j is the expected return time to state j.

Exercises

3.6.1 Find an invariant distribution λ for the Q-matrix

$$Q = \begin{pmatrix} -2 & 1 & 1 \\ 4 & -4 & 0 \\ 2 & 1 & -3 \end{pmatrix}$$

and verify that $\lim_{t\to\infty} p_{11}(t) = \lambda_1$ using your answer to Exercise 2.1.1.

3.6.2 In each of the following cases, compute $\lim_{t\to\infty} \mathbb{P}(X_t = 2 | X_0 = 1)$ for the Markov chain $(X_t)_{t\geq 0}$ with the given Q-matrix on $\{1, 2, 3, 4\}$:

(a) $\begin{pmatrix} -2 & 1 & 1 & 0 \\ 0 & -1 & 1 & 0 \\ 0 & 0 & -1 & 1 \\ 1 & 0 & 0 & -1 \end{pmatrix}$ (b) $\begin{pmatrix} -2 & 1 & 1 & 0 \\ 0 & -1 & 1 & 0 \\ 0 & 0 & -1 & 1 \\ 0 & 0 & 0 & 0 \end{pmatrix}$

(c) $\begin{pmatrix} -1 & 1 & 0 & 0 \\ 1 & -1 & 0 & 0 \\ 0 & 0 & -2 & 2 \\ 0 & 0 & 2 & -2 \end{pmatrix}$ (d) $\begin{pmatrix} -2 & 1 & 0 & 1 \\ 0 & -2 & 2 & 0 \\ 0 & 1 & -1 & 0 \\ 0 & 0 & 0 & 0 \end{pmatrix}$

3.6.3 Customers arrive at a single-server queue in a Poisson stream of rate λ. Each customer has a service requirement distributed as the sum of two independent exponential random variables of parameter μ. Service requirements are independent of one another and of the arrival process. Write down the generator matrix Q of a continuous-time Markov chain which models this, explaining what the states of the chain represent. Calculate the essentially unique invariant measure for Q, and deduce that the chain is positive recurrent if and only if $\lambda/\mu < 1/2$.

3.7 Time reversal

Time reversal of continuous-time chains has the same features found in the discrete-time case. Reversibility provides a powerful tool in the analysis of Markov chains, as we shall see in Section 5.2. Note in the following result how time reversal interchanges the roles of backward and forward equations. This echoes our proof of the forward equation, which rested on the time reversal identity of Lemma 2.8.5.

A small technical point arises in time reversal: right-continuous processes become left-continuous processes. For the processes we consider, this is unimportant. We could if we wished redefine the time-reversed process to equal its right limit at the jump times, thus obtaining again a right-continuous process. We shall suppose implicitly that this is done, and forget about the problem.

Theorem 3.7.1. *Let Q be irreducible and non-explosive and suppose that Q has an invariant distribution λ. Let $T \in (0, \infty)$ be given and let $(X_t)_{0 \leq t \leq T}$ be Markov(λ, Q). Set $\widehat{X}_t = X_{T-t}$. Then the process $(\widehat{X}_t)_{0 \leq t \leq T}$ is Markov$(\lambda, \widehat{Q})$, where $\widehat{Q} = (\widehat{q}_{ij} : i, j \in I)$ is given by $\lambda_j \widehat{q}_{ji} = \lambda_i q_{ij}$. Moreover, $\widehat{Q}$ is also irreducible and non-explosive with invariant distribution λ.*

Proof. By Theorem 2.8.6, the semigroup $(P(t) : t \geq 0)$ of Q is the minimal non-negative solution of the forward equation
$$P'(t) = P(t)Q, \quad P(0) = I.$$
Also, for all $t > 0$, $P(t)$ is an irreducible stochastic matrix with invariant distribution λ. Define $\widehat{P}(t)$ by
$$\lambda_j \widehat{p}_{ji}(t) = \lambda_i p_{ij}(t),$$
then $\widehat{P}(t)$ is an irreducible stochastic matrix with invariant distribution λ, and we can rewrite the forward equation transposed as
$$\widehat{P}'(t) = \widehat{Q}\widehat{P}(t).$$
But this is the backward equation for $\widehat{Q}$, which is itself a Q-matrix, and $\widehat{P}(t)$ is then its minimal non-negative solution. Hence $\widehat{Q}$ is irreducible and non-explosive and has invariant distribution λ.

Finally, for $0 = t_0 < \ldots < t_n = T$ and $s_k = t_k - t_{k-1}$, by Theorem 2.8.4 we have
$$\mathbb{P}(\widehat{X}_{t_0} = i_0, \ldots, \widehat{X}_{t_n} = i_n) = \mathbb{P}(X_{T-t_0} = i_0, \ldots, X_{T-t_n} = i_n)$$
$$= \lambda_{i_n} p_{i_n i_{n-1}}(s_n) \ldots p_{i_1 i_0}(s_1)$$
$$= \lambda_{i_0} \widehat{p}_{i_0 i_1}(s_1) \ldots \widehat{p}_{i_{n-1} i_n}(s_n)$$
so, by Theorem 2.8.4 again, $(\widehat{X}_t)_{0 \leq t \leq T}$ is Markov$(\lambda, \widehat{Q})$. □

The chain $(\widehat{X}_t)_{0 \leq t \leq T}$ is called the *time-reversal* of $(X_t)_{0 \leq t \leq T}$.

A Q-matrix Q and a measure λ are said to be in *detailed balance* if
$$\lambda_i q_{ij} = \lambda_j q_{ji} \qquad \text{for all } i, j.$$

Lemma 3.7.2. *If Q and λ are in detailed balance then λ is invariant for Q.*

Proof. We have $(\lambda Q)_i = \sum_{j \in I} \lambda_j q_{ji} = \sum_{j \in I} \lambda_i q_{ij} = 0.$ □

Let $(X_t)_{t \geq 0}$ be Markov(λ, Q), with Q irreducible and non-explosive. We say that $(X_t)_{t \geq 0}$ is *reversible* if, for all $T > 0$, $(X_{T-t})_{0 \leq t \leq T}$ is also Markov(λ, Q).

Theorem 3.7.3. *Let Q be an irreducible and non-explosive Q-matrix and let λ be a distribution. Suppose that $(X_t)_{t \geq 0}$ is Markov(λ, Q). Then the following are equivalent:*
  (a) *$(X_t)_{t \geq 0}$ is reversible;*
  (b) *Q and λ are in detailed balance.*

Proof. Both (a) and (b) imply that λ is invariant for Q. Then both (a) and (b) are equivalent to the statement that $\widehat{Q} = Q$ in Theorem 3.7.1. □

Exercise

3.7.1 Consider a fleet of N buses. Each bus breaks down independently at rate μ, when it is sent to the depot for repair. The repair shop can only repair one bus at a time and each bus takes an exponential time of parameter λ to repair. Find the equilibrium distribution of the number of buses in service.

3.7.2 Calls arrive at a telephone exchange as a Poisson process of rate λ, and the lengths of calls are independent exponential random variables of parameter μ. Assuming that infinitely many telephone lines are available, set up a Markov chain model for this process.

Show that for large t the distribution of the number of lines in use at time t is approximately Poisson with mean λ/μ.

Find the mean length of the busy periods during which at least one line is in use.

Show that the expected number of lines in use at time t, given that n are in use at time 0, is $ne^{-\mu t} + \lambda(1 - e^{-\mu t})/\mu$.

Show that, in equilibrium, the number N_t of calls finishing in the time interval $[0, t]$ has Poisson distribution of mean λt.

Is $(N_t)_{t \geq 0}$ a Poisson process?

3.8 Ergodic theorem

Long-run averages for continuous-time chains display the same sort of behaviour as in the discrete-time case, and for similar reasons. Here is the result.

Theorem 3.8.1 (Ergodic theorem). Let Q be irreducible and let ν be any distribution. If $(X_t)_{t\geq 0}$ is Markov(ν, Q), then

$$\mathbb{P}\left(\frac{1}{t}\int_0^t \mathbf{1}_{\{X_s=i\}}ds \to \frac{1}{m_i q_i} \text{ as } t \to \infty\right) = 1$$

where $m_i = \mathbb{E}_i(T_i)$ is the expected return time to state i. Moreover, in the positive recurrent case, for any bounded function $f: I \to \mathbb{R}$ we have

$$\mathbb{P}\left(\frac{1}{t}\int_0^t f(X_s)ds \to \overline{f} \text{ as } t \to \infty\right) = 1$$

where

$$\overline{f} = \sum_{i \in I} \lambda_i f_i$$

and where $(\lambda_i : i \in I)$ is the unique invariant distribution.

Proof. If Q is transient then the total time spent in any state i is finite, so

$$\frac{1}{t}\int_0^t \mathbf{1}_{\{X_s=i\}}ds \leq \frac{1}{t}\int_0^\infty \mathbf{1}_{\{X_s=i\}}ds \to 0 = \frac{1}{m_i}.$$

Suppose then that Q is recurrent and fix a state i. Then $(X_t)_{t\geq 0}$ hits i with probability 1 and the long-run proportion of time in i equals the long-run proportion of time in i after first hitting i. So, by the strong Markov property (of the jump chain), it suffices to consider the case $\nu = \delta_i$.

Denote by M_i^n the length of the nth visit to i, by T_i^n the time of the nth return to i and by L_i^n the length of the nth excursion to i. Thus for $n = 0, 1, 2, \ldots$, setting $T_i^0 = 0$, we have

$$M_i^{n+1} = \inf\{t > T_i^n : X_t \neq i\} - T_i^n$$
$$T_i^{n+1} = \inf\{t > T_i^n + M_i^{n+1} : X_t = i\}$$
$$L_i^{n+1} = T_i^{n+1} - T_i^n.$$

By the strong Markov property (of the jump chain) at the stopping times T_i^n for $n \geq 0$ we find that $L_i^1, L_i^2, \ldots$ are independent and identically distributed with mean m_i, and that $M_i^1, M_i^2, \ldots$ are independent and identically distributed with mean $1/q_i$. Hence, by the strong law of large numbers (see Theorem 1.10.1)

$$\frac{L_i^1 + \cdots + L_i^n}{n} \to m_i \text{ as } n \to \infty$$
$$\frac{M_i^1 + \cdots + M_i^n}{n} \to \frac{1}{q_i} \text{ as } n \to \infty$$

3.8 Ergodic theorem

and hence

$$\frac{M_i^1 + \cdots + M_i^n}{L_i^1 + \cdots + L_i^n} \to \frac{1}{m_i q_i} \quad \text{as } n \to \infty$$

with probability 1. In particular, we note that $T_i^n/T_i^{n+1} \to 1$ as $n \to \infty$ with probability 1. Now, for $T_i^n \leq t < T_i^{n+1}$ we have

$$\frac{T_i^n}{T_i^{n+1}} \frac{M_i^1 + \cdots + M_i^n}{L_i^1 + \cdots + L_i^n} \leq \frac{1}{t}\int_0^t 1_{\{X_s = i\}} ds \leq \frac{T_i^{n+1}}{T_i^n} \frac{M_i^1 + \cdots + M_i^{n+1}}{L_i^1 + \cdots + L_i^{n+1}}$$

so on letting $t \to \infty$ we have, with probability 1

$$\frac{1}{t}\int_0^t 1_{\{X_s = i\}} ds \to \frac{1}{m_i q_i}.$$

In the positive recurrent case we can write

$$\frac{1}{t}\int_0^t f(X_s)ds - \overline{f} = \sum_{i \in I} f_i \left(\frac{1}{t}\int_0^t 1_{\{X_s = i\}} ds - \lambda_i\right)$$

where $\lambda_i = 1/(m_i q_i)$. We conclude that

$$\frac{1}{t}\int_0^t f(X_s)ds \to \overline{f} \quad \text{as } t \to \infty$$

with probability 1, by the same argument as was used in the proof of Theorem 1.10.2. □

4
Further theory

In the first three chapters we have given an account of the elementary theory of Markov chains. This already covers a great many applications, but is just the beginning of the theory of Markov processes. The further theory inevitably involves more sophisticated techniques which, although having their own interest, can obscure the overall structure. On the other hand, the overall structure is, to a large extent, already present in the elementary theory. We therefore thought it worth while to discuss some features of the further theory in the context of simple Markov chains, namely, martingales, potential theory, electrical networks and Brownian motion. The idea is that the Markov chain case serves as a guiding metaphor for more complicated processes. So the reader familiar with Markov chains may find this chapter helpful alongside more general higher-level texts. At the same time, further insight is gained into Markov chains themselves.

4.1 Martingales

A martingale is a process whose average value remains constant in a particular strong sense, which we shall make precise shortly. This is a sort of balancing property. Often, the identification of martingales is a crucial step in understanding the evolution of a stochastic process.

We begin with a simple example. Consider the simple symmetric random walk $(X_n)_{n\geq 0}$ on $\mathbb{Z}$, which is a Markov chain with the following diagram

The average value of the walk is constant; indeed it has the stronger property that the average value of the walk at some future time is always simply the current value. In precise terms we have

$$\mathbb{E}X_n = \mathbb{E}X_0;$$

and the stronger property says that, for $n \geq m$,

$$\mathbb{E}(X_n - X_m \mid X_0 = i_0, \ldots, X_m = i_m) = 0.$$

This stronger property says that $(X_n)_{n \geq 0}$ is in fact a martingale.

Here is the general definition. Let us fix for definiteness a Markov chain $(X_n)_{n \geq 0}$ and write $\mathcal{F}_n$ for the collection of all sets depending only on $X_0, \ldots, X_n$. The sequence $(\mathcal{F}_n)_{n \geq 0}$ is called the *filtration* of $(X_n)_{n \geq 0}$ and we think of $\mathcal{F}_n$ as representing the state of knowledge, or history, of the chain up to time n. A process $(M_n)_{n \geq 0}$ is called *adapted* if M_n depends only on $X_0, \ldots, X_n$. A process $(M_n)_{n \geq 0}$ is called *integrable* if $\mathbb{E}|M_n| < \infty$ for all n. An adapted integrable process $(M_n)_{n \geq 0}$ is called a *martingale* if

$$\mathbb{E}[(M_{n+1} - M_n)1_A] = 0$$

for all $A \in \mathcal{F}_n$ and all n. Since the collection $\mathcal{F}_n$ consists of countable unions of elementary events such as

$$\{X_0 = i_0, X_1 = i_1, \ldots, X_n = i_n\},$$

this martingale property is equivalent to saying that

$$\mathbb{E}(M_{n+1} - M_n \mid X_0 = i_0, \ldots, X_n = i_n) = 0$$

for all $i_0, \ldots, i_n$ and all n.

A third formulation of the martingale property involves another notion of conditional expectation. Given an integrable random variable Y, we define

$$\mathbb{E}(Y \mid \mathcal{F}_n) = \sum_{i_0, \ldots, i_n} \mathbb{E}(Y \mid X_0 = i_0, \ldots, X_n = i_n) 1_{\{X_0 = i_0, \ldots, X_n = i_n\}}.$$

The random variable $\mathbb{E}(Y \mid \mathcal{F}_n)$ is called the *conditional expectation* of Y given $\mathcal{F}_n$. In passing from Y to $\mathbb{E}(Y \mid \mathcal{F}_n)$, what we do is to replace on each elementary event $A \in \mathcal{F}_n$, the random variable Y by its average value $\mathbb{E}(Y \mid A)$. It is easy to check that an adapted integrable process $(M_n)_{n \geq 0}$ is a martingale if and only if

$$\mathbb{E}(M_{n+1} \mid \mathcal{F}_n) = M_n \quad \text{for all } n.$$

Conditional expectation is a partial averaging, so if we complete the process and average the conditional expectation we should get the full expectation

$$\mathbb{E}\big(\mathbb{E}(Y \mid \mathcal{F}_n)\big) = \mathbb{E}(Y).$$

It is easy to check that this formula holds.

In particular, for a martingale

$$\mathbb{E}(M_n) = \mathbb{E}\big(\mathbb{E}(M_{n+1} \mid \mathcal{F}_n)\big) = \mathbb{E}(M_{n+1})$$

so, by induction

$$\mathbb{E}(M_n) = \mathbb{E}(M_0).$$

This was already clear on taking $A = \Omega$ in our original definition of a martingale.

We shall prove one general result about martingales, then see how it explains some things we know about the simple symmetric random walk. Recall that a random variable

$$T : \Omega \to \{0, 1, 2, \dots\} \cup \{\infty\}$$

is a *stopping time* if $\{T = n\} \in \mathcal{F}_n$ for all $n < \infty$. An equivalent condition is that $\{T \leq n\} \in \mathcal{F}_n$ for all $n < \infty$. Recall from Section 1.4 that all sorts of hitting times are stopping times.

Theorem 4.1.1 (Optional stopping theorem). *Let $(M_n)_{n \geq 0}$ be a martingale and let T be a stopping time. Suppose that at least one of the following conditions holds:*

(i) *$T \leq n$ for some n;*

(ii) *$T < \infty$ and $|M_n| \leq C$ whenever $n \leq T$.*

Then $\mathbb{E} M_T = \mathbb{E} M_0$.

Proof. Assume that (i) holds. Then

$$M_T - M_0 = (M_T - M_{T-1}) + \ldots + (M_1 - M_0)$$
$$= \sum_{k=0}^{n-1} (M_{k+1} - M_k) 1_{k < T}.$$

Now $\{k < T\} = \{T \leq k\}^c \in \mathcal{F}_k$ since T is a stopping time, and so

$$\mathbb{E}[(M_{k+1} - M_k)1_{k<T}] = 0$$

since $(M_k)_{k \geq 0}$ is a martingale. Hence

$$\mathbb{E}M_T - \mathbb{E}M_0 = \sum_{k=0}^{n-1} \mathbb{E}[(M_{k+1} - M_k)1_{k<T}] = 0.$$

If we do not assume (i) but (ii), then the preceding argument applies to the stopping time $T \wedge n$, so that $\mathbb{E}M_{T \wedge n} = \mathbb{E}M_0$. Then

$$|\mathbb{E}M_T - \mathbb{E}M_0| = |\mathbb{E}M_T - \mathbb{E}M_{T \wedge n}| \leq \mathbb{E}|M_T - M_{T \wedge n}| \leq 2C\mathbb{P}(T > n)$$

for all n. But $\mathbb{P}(T > n) \to 0$ as $n \to \infty$, so $\mathbb{E}M_T = \mathbb{E}M_0$. □

Returning to the simple symmetric random walk $(X_n)_{n \geq 0}$, suppose that $X_0 = 0$ and we take

$$T = \inf\{n \geq 0 : X_n = -a \quad \text{or} \quad X_n = b\}$$

where $a, b \in \mathbb{N}$ are given. Then T is a stopping time and $T < \infty$ by recurrence of finite closed classes. Thus condition (ii) of the optional stopping theorem applies with $M_n = X_n$ and $C = a \vee b$. We deduce that $\mathbb{E}X_T = \mathbb{E}X_0 = 0$. So what? Well, now we can compute

$$p = \mathbb{P}(X_n \text{ hits } -a \text{ before } b).$$

We have $X_T = -a$ with probability p and $X_T = b$ with probability $1-p$, so

$$0 = \mathbb{E}X_T = p(-a) + (1-p)b$$

giving

$$p = b/(a+b).$$

There is an entirely different, Markovian, way to compute p, using the methods of Section 1.4. But the intuition behind the result $\mathbb{E}X_T = 0$ is very clear: a gambler, playing a fair game, leaves the casino once losses reach a or winnings reach b, whichever is sooner; since the game is fair, the average gain should be zero.

We discussed in Section 1.3 the counter-intuitive case of a gambler who keeps on playing a fair game against an infinitely rich casino, with the certain outcome of ruin. This game ends at the finite stopping time

$$T = \inf\{n \geq 0 : X_n = -a\}$$

where a is the gambler's initial fortune. Since $X_T = -a$ we have

$$\mathbb{E}X_T = -a \neq 0 = \mathbb{E}X_0$$

but this does not contradict the optional stopping theorem because neither condition (i) nor condition (ii) is satisfied. Thus, while intuition might suggest that $\mathbb{E}X_T = \mathbb{E}X_0$ is rather obvious, some care is needed as it is not always true.

The example just discussed was rather special in that the chain $(X_n)_{n\geq 0}$ itself was a martingale. Obviously, this is not true in general; indeed a martingale is necessarily real-valued and we do not in general insist that the state-space I is contained in $\mathbb{R}$. Nevertheless, to every Markov chain is associated a whole collection of martingales, and these martingales characterize the chain. This is the basis of a deep connection between martingales and Markov chains.

We recall that, given a function $f : I \to \mathbb{R}$ and a Markov chain $(X_n)_{n\geq 0}$ with transition matrix P, we have

$$(P^n f)(i) = \sum_{j\in I} p_{ij}^{(n)} f_j = \mathbb{E}_i\big(f(X_n)\big).$$

Theorem 4.1.2. *Let $(X_n)_{n\geq 0}$ be a random process with values in I and let P be a stochastic matrix. Write $(\mathcal{F}_n)_{n\geq 0}$ for the filtration of $(X_n)_{n\geq 0}$. Then the following are equivalent:*

(i) *$(X_n)_{n\geq 0}$ is a Markov chain with transition matrix P;*

(ii) *for all bounded functions $f : I \to \mathbb{R}$, the following process is a martingale:*

$$M_n^f = f(X_n) - f(X_0) - \sum_{m=0}^{n-1}(P - I)f(X_m).$$

Proof. Suppose (i) holds. Let f be a bounded function. Then

$$|(Pf)(i)| = |\sum_{j\in I} p_{ij} f_j| \leq \sup_j |f_j|$$

so

$$|M_n^f| \leq 2(n+1)\sup_j |f_j| < \infty$$

showing that M_n^f is integrable for all n.

Let $A = \{X_0 = i_0, \ldots, X_n = i_n\}$. By the Markov property

$$\mathbb{E}\big(f(X_{n+1}) \mid A\big) = \mathbb{E}_{i_n}\big(f(X_1)\big) = (Pf)(i_n)$$

so
$$\mathbb{E}(M_{n+1}^f - M_n^f \mid A) = \mathbb{E}[f(X_{n+1}) - (Pf)(X_n) \mid A] = 0$$
and so $(M_n^f)_{n \geq 0}$ is a martingale.

On the other hand, if (ii) holds, then
$$\mathbb{E}[f(X_{n+1}) - (Pf)(X_n) \mid X_0 = i_0, \ldots, X_n = i_n] = 0$$
for all bounded functions f. On taking $f = 1_{\{i_{n+1}\}}$ we obtain
$$\mathbb{P}(X_{n+1} = i_{n+1} \mid X_0 = i_0, \ldots, X_n = i_n) = p_{i_n i_{n+1}}$$
so $(X_n)_{n \geq 0}$ is Markov with transition matrix P. □

Some more martingales associated to a Markov chain are described in the next result. Notice that we drop the requirement that f be bounded.

Theorem 4.1.3. *Let $(X_n)_{n \geq 0}$ be a Markov chain with transition matrix P. Suppose that a function $f : \mathbb{Z}^+ \times I \to \mathbb{R}$ satisfies, for all $n \geq 0$, both*
$$\mathbb{E}|f(n, X_n)| < \infty$$
and
$$(Pf)(n+1, i) = \sum_{j \in I} p_{ij} f(n+1, j) = f(n, i).$$
Then $M_n = f(n, X_n)$ is a martingale.

Proof. We have assumed that M_n is integrable for all n. Then, by the Markov property
$$\mathbb{E}(M_{n+1} - M_n \mid X_0 = i_0, \ldots, X_n = i_n) = \mathbb{E}_{i_n}[f(n+1, X_1) - f(n, X_0)]$$
$$= (Pf)(n+1, i_n) - f(n, i_n) = 0.$$
So $(M_n)_{n \geq 0}$ is a martingale. □

Let us see how this theorem works in the case where $(X_n)_{n \geq 0}$ is a simple random walk on $\mathbb{Z}$, starting from 0. We consider $f(i) = i$ and $g(n, i) = i^2 - n$. Since $|X_n| \leq n$ for all n, we have
$$\mathbb{E}|f(X_n)| < \infty, \quad \mathbb{E}|g(n, X_n)| < \infty.$$
Also
$$(Pf)(i) = (i-1)/2 + (i+1)/2 = i = f(i),$$
$$(Pg)(n+1, i) = (i-1)^2/2 + (i+1)^2/2 - (n+1) = i^2 - n = g(n, i).$$
Hence both $X_n = f(X_n)$ and $Y_n = g(n, X_n)$ are martingales.

In order to put this to some use, consider again the stopping time
$$T = \inf\{n \geq 0 : X_n = -a \quad \text{or} \quad X_n = b\}$$
where $a, b \in \mathbb{N}$. By the optional stopping theorem
$$0 = \mathbb{E}(Y_0) = \mathbb{E}(Y_{T \wedge n}) = \mathbb{E}(X^2_{T \wedge n}) - \mathbb{E}(T \wedge n).$$
Hence
$$\mathbb{E}(T \wedge n) = \mathbb{E}(X^2_{T \wedge n}).$$
On letting $n \to \infty$, the left side converges to $\mathbb{E}(T)$, by monotone convergence, and the right side to $\mathbb{E}(X^2_T)$ by bounded convergence. So we obtain
$$\mathbb{E}(T) = \mathbb{E}(X^2_T) = a^2 p + b^2(1-p) = ab.$$

We have given only the simplest examples of the use of martingales in studying Markov chains. Some more will appear in later sections. For an excellent introduction to martingales and their applications we recommend *Probability with Martingales* by David Williams (Cambridge University Press, 1991).

Exercise

4.1.1 Let $(X_n)_{n \geq 0}$ be a Markov chain on I and let A be an absorbing set in I. Set
$$T = \inf\{n \geq 0 : X_n \in A\}$$
and
$$h_i = \mathbb{P}_i(X_n \in A \text{ for some } n \geq 0) = \mathbb{P}_i(T < \infty).$$
Show that $M_n = h(X_n)$ is a martingale.

4.2 Potential theory

Several physical theories share a common mathematical framework, which is known as potential theory. One example is Newton's theory of gravity, but potential theory is also relevant to electrostatics, fluid flow and the diffusion of heat. In gravity, a distribution of mass, of density ρ say, gives rise to a gravitational potential ϕ, which in suitable units satisfies the equation
$$-\Delta\phi = \rho,$$
where $\Delta = \partial^2/\partial x^2 + \partial^2/\partial y^2 + \partial^2/\partial z^2$. The potential ϕ is felt physically through its gradient
$$\nabla\phi = \left(\frac{\partial\phi}{\partial x}, \frac{\partial\phi}{\partial y}, \frac{\partial\phi}{\partial z}\right)$$

4.2 Potential theory

which gives the force of gravity acting on a particle of unit mass. Markov chains, where space is discrete, obviously have no direct link with this theory, in which space is a continuum. An indirect link is provided by Brownian motion, which we shall discuss in Section 4.4.

In this section we are going to consider *potential theory for a countable state-space*, which has much of the structure of the continuum version. This discrete theory amounts to doing Markov chains without the probability, which has the disadvantage that one loses the intuitive picture of the process, but the advantage of wider applicability. We shall begin by introducing the idea of potentials associated to a Markov chain, and by showing how to calculate these potentials. This is a unifying idea, containing within it other notions previously considered such as hitting probabilities and expected hitting times. It also finds application when one associates costs to Markov chains in modelling economic activity: see Section 5.4.

Once we have established the basic link between a Markov chain and its associated potentials, we shall briefly run through some of the main features of potential theory, explaining their significance in terms of Markov chains. This is the easiest way to appreciate the general structure of potential theory, unobscured by technical difficulties. The basic ideas of boundary theory for Markov chains will also be introduced.

Before we embark on a general discussion of potentials associated to a Markov chain, here are two simple examples. In these examples the potential ϕ has the interpretation of expected total cost.

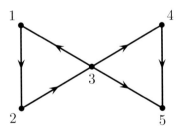

Example 4.2.1

Consider the discrete-time random walk on the directed graph shown above, which at each step choses among the allowable transitions with equal probability. Suppose that on each visit to states $i = 1, 2, 3, 4$ a cost c_i is incurred, where $c_i = i$. What is the fair price to move from state 3 to state 4?

The fair price is always the difference in the expected total cost. We denote by ϕ_i the expected total cost starting from i. Obviously, $\phi_5 = 0$ and by considering the effect of a single step we see that

$$\phi_1 = 1 + \phi_2,$$
$$\phi_2 = 2 + \phi_3,$$
$$\phi_3 = 3 + \tfrac{1}{3}\phi_1 + \tfrac{1}{3}\phi_4,$$
$$\phi_4 = 4.$$

Hence $\phi_3 = 8$ and the fair price to move from 3 to 4 is 4.

We shall now consider two variations on this problem. First suppose our process is, instead, the continuous-time random walk $(X_t)_{t \geq 0}$ on the same directed graph which makes each allowable transition at rate 1, and suppose cost is incurred at rate $c_i = i$ in state i for $i = 1, 2, 3, 4$. Thus the total cost is now

$$\int_0^\infty c(X_s)ds.$$

What now is the fair price to move from 3 to 4? The expected cost incurred on each visit to i is given by c_i/q_i and $q_1 = 1, q_2 = 1, q_3 = 3, q_4 = 1$. So we see, as before

$$\phi_1 = 1 + \phi_2,$$
$$\phi_2 = 2 + \phi_3,$$
$$\phi_3 = \tfrac{3}{3} + \tfrac{1}{3}\phi_1 + \tfrac{1}{3}\phi_4,$$
$$\phi_4 = 4.$$

Hence $\phi_3 = 5$ and the fair price to move from 3 to 4 is 1.

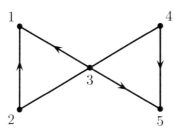

In the second variation we consider the discrete-time random walk $(X_n)_{n \geq 0}$ on the modified graph shown above. Where there is no arrow, transitions are allowed in both directions. Obviously, states 1 and 5 are absorbing. We impose a cost $c_i = i$ on each visit to i for $i = 2, 3, 4$, and a final cost f_i on arrival at $i = 1$ or 5, where $f_i = i$. Thus the total cost is now

$$\sum_{n=0}^{T-1} c(X_n) + f(X_T)$$

where T is the hitting time of $\{1,5\}$. Write, as before, ϕ_i for the expected total cost starting from i. Then $\phi_1 = 1$, $\phi_5 = 5$ and

$$\phi_2 = 2 + \tfrac{1}{2}(\phi_1 + \phi_3),$$
$$\phi_3 = 3 + \tfrac{1}{4}(\phi_1 + \phi_2 + \phi_4 + \phi_5),$$
$$\phi_4 = 4 + \tfrac{1}{2}(\phi_3 + \phi_5).$$

On solving these equations we obtain $\phi_2 = 7$, $\phi_3 = 9$ and $\phi_4 = 11$. So in this case the fair price to move from 3 to 4 is -2.

Example 4.2.2

Consider the simple discrete-time random walk on $\mathbb{Z}$ with transition probabilities $p_{i,i-1} = q < p = p_{i,i+1}$. Let $c > 0$ and suppose that a cost c^i is incurred every time the walk visits state i. What is the expected total cost ϕ_0 incurred by the walk starting from 0?

We must be prepared to find that $\phi_0 = \infty$ for some values of c, as the total cost is a sum over infinitely many times. Indeed, we know that the walk $X_n \to \infty$ with probability 1, so for $c \geq 1$ we shall certainly have $\phi_0 = \infty$.

Let ϕ_i denote the expected total cost starting from i. On moving one step to the right, all costs are multiplied by c, so we must have

$$\phi_{i+1} = c\phi_i.$$

By considering what happens on the first step, we see

$$\phi_0 = 1 + p\phi_1 + q\phi_{-1} = 1 + (cp + q/c)\phi_0.$$

Note that $\phi_0 = \infty$ always satisfies this equation. We shall see in the general theory that ϕ_0 is the minimal non-negative solution. Let us look for a finite solution: then

$$-(c^2 p - c + q)\phi_0 = c$$

so

$$\phi_0 = \frac{c}{c - c^2 p - q}.$$

The quadratic $c^2 p - c + q$ has roots at q/p and 1, and takes negative values in between. Hence the expected total cost is given by

$$\phi_0 = \begin{cases} c/(c - c^2 p - q) & \text{if } c \in (q/p, 1) \\ \infty & \text{otherwise.} \end{cases}$$

It was clear at the outset that $\phi_0 = \infty$ when $c \geq 1$. It is interesting that $\phi_0 = \infty$ also when c is too small: in this case the costs rapidly become large to the left of 0, and although the walk eventually drifts away to the right, the expected cost incurred to the left of 0 is infinite.

In the examples just discussed we were able to calculate potentials by writing down and solving a system of linear equations. This situation is familiar from hitting probabilities and expected hitting times. Indeed, these are simple examples of potentials for Markov chains. As the examples show, one does not really need a general theory to write down the linear equations. Nevertheless, we are now going to give some general results on potentials. These will help to reveal the scope of the ideas used in the examples, and will reveal also what happens when the linear equations do not have a unique solution. We shall discuss the cases of discrete and continuous time side-by-side. Throughout, we shall write $(X_n)_{n\geq 0}$ for a discrete-time chain with transition matrix P, and $(X_t)_{t\geq 0}$ for a continuous-time chain with generator matrix Q. As usual, we insist that $(X_t)_{t\geq 0}$ be minimal.

Let us partition the state-space I into two disjoint sets D and ∂D; we call ∂D the *boundary*. We suppose that functions $(c_i : i \in D)$ and $(f_i : i \in \partial D)$ are given. We shall consider the associated *potential*, defined by

$$\phi_i = \mathbb{E}_i \left(\sum_{n<T} c(X_n) + f(X_T) 1_{T<\infty} \right)$$

in discrete time, and in continuous time

$$\phi_i = \mathbb{E}_i \left(\int_0^T c(X_t) dt + f(X_T) 1_{T<\infty} \right),$$

where T denotes the hitting time of ∂D. To be sure that the sums and integrals here are well defined, we shall assume for the most part that c and f are *non-negative*, that is, $c_i \geq 0$ for all $i \in D$ and $f_i \geq 0$ for all $i \in \partial D$. More generally, ϕ is the difference of the potentials associated with the positive and negative parts of c and f, so this assumption is not too restrictive. In the explosive case we always set $c(\infty) = 0$, so no further costs are incurred after explosion.

The most obvious interpretation of these potentials is in terms of cost: the chain wanders around in D until it hits the boundary: whilst in D, at state i say, it incurs a *cost* c_i per unit time; when and if it hits the boundary, at j say, a *final cost* f_j is incurred. Note that we do not assume the chain will hit the boundary, or even that the boundary is non-empty.

Theorem 4.2.3. Suppose that $(c_i : i \in D)$ and $(f_i : i \in \partial D)$ are non-negative. Set
$$\phi_i = \mathbb{E}_i \left(\sum_{n<T} c(X_n) + f(X_T) 1_{T<\infty} \right)$$
where T denotes the hitting time of ∂D. Then

(i) the potential $\phi = (\phi_i : i \in I)$ satisfies
$$\begin{cases} \phi = P\phi + c & \text{in } D \\ \phi = f & \text{in } \partial D; \end{cases} \qquad (4.1)$$

(ii) if $\psi = (\psi_i : i \in I)$ satisfies
$$\begin{cases} \psi \geq P\psi + c & \text{in } D \\ \psi \geq f & \text{in } \partial D \end{cases} \qquad (4.2)$$
and $\psi_i \geq 0$ for all i, then $\psi_i \geq \phi_i$ for all i;

(iii) if $\mathbb{P}_i(T < \infty) = 1$ for all i, then (4.1) has at most one bounded solution.

Proof. (i) Obviously, $\phi = f$ on ∂D. For $i \in D$ by the Markov property
$$\mathbb{E}_i \left(\sum_{1 \leq n<T} c(X_n) + f(X_T) 1_{T<\infty} \Big| X_1 = j \right)$$
$$= \mathbb{E}_j \left(\sum_{n<T} c(X_n) + f(X_T) 1_{T<\infty} \right) = \phi_j$$
so we have
$$\phi_i = c_i + \sum_{j \in I} p_{ij} \mathbb{E} \left(\sum_{1 \leq n<T} c(X_n) + f(X_T) 1_{T<\infty} \Big| X_1 = j \right)$$
$$= c_i + \sum_{j \in I} p_{ij} \phi_j$$
as required.

(ii) Consider the expected cost up to time n:
$$\phi_i(n) = \mathbb{E}_i \left(\sum_{k=0}^n c(X_k) 1_{k<T} + f(X_T) 1_{T \leq n} \right).$$

By monotone convergence, $\phi_i(n) \uparrow \phi_i$ as $n \to \infty$. Also, by the argument used in part (i), we find

$$\begin{cases} \phi(n+1) = c + P\phi(n) & \text{in } D \\ \phi(n+1) = f & \text{in } \partial D. \end{cases}$$

Suppose that ψ satisfies (4.2) and $\psi \geq 0 = \phi(0)$. Then $\psi \geq P\psi + c \geq P\phi(0) + c = \phi(1)$ in D and $\psi \geq f = \phi(1)$ in ∂D, so $\psi \geq \phi(1)$. Similarly and by induction, $\psi \geq \phi(n)$ for all n, and hence $\psi \geq \phi$.

(iii) We shall show that if ψ satisfies (4.2) then

$$\psi_i \geq \phi_i(n-1) + \mathbb{E}_i\big(\psi(X_n)1_{T \geq n}\big),$$

with equality if equality holds in (4.2). This is another proof of (ii). But also, in the case of equality, if $|\psi_i| \leq M$ and $\mathbb{P}_i(T < \infty) = 1$ for all i, then as $n \to \infty$

$$|\mathbb{E}_i\big(\psi(X_n)1_{T \geq n}\big)| \leq M\mathbb{P}_i(T \geq n) \to 0$$

so $\psi = \lim_{n \to \infty} \phi(n) = \phi$, proving (iii).

For $i \in D$ we have

$$\psi_i \geq c_i + \sum_{j \in \partial D} p_{ij} f_j + \sum_{j \in D} p_{ij} \psi_j$$

and, by repeated substitution for ψ on the right

$$\psi_i \geq c_i + \sum_{j \in \partial D} p_{ij} f_j + \sum_{j \in D} p_{ij} c_j$$

$$+ \ldots + \sum_{j_1 \in D} \cdots \sum_{j_{n-1} \in D} p_{ij_1} \cdots p_{j_{n-2} j_{n-1}} c_{j_{n-1}}$$

$$+ \sum_{j_1 \in D} \cdots \sum_{j_{n-1} \in D} \sum_{j_n \in \partial D} p_{ij_1} \cdots p_{j_{n-1} j_n} f_{j_n}$$

$$+ \sum_{j_1 \in D} \cdots \sum_{j_n \in D} p_{ij_1} \cdots p_{j_{n-1} j_n} \psi_{j_n}$$

$$= \mathbb{E}_i\Big(c(X_0)1_{T>0} + f(X_1)1_{T=1} + c(X_1)1_{T>1}$$

$$+ \ldots + c(X_{n-1})1_{T>n-1} + f(X_n)1_{T=n} + \psi(X_n)1_{T>n}\Big)$$

$$= \phi_i(n-1) + \mathbb{E}_i\big(\psi(X_n)1_{T \geq n}\big)$$

as required, with equality when equality holds in (4.2). □

4.2 Potential theory

It is illuminating to think of the calculation we have just done in terms of martingales. Consider

$$M_n = \sum_{k=0}^{n-1} c(X_k)1_{k<T} + f(X_T)1_{T<n} + \psi(X_n)1_{n\leq T}.$$

Then

$$\mathbb{E}(M_{n+1} \mid \mathcal{F}_n) = \sum_{k=0}^{n-1} c(X_k)1_{k<T} + f(X_T)1_{T<n}$$
$$+ (P\psi + c)(X_n)1_{T>n} + f(X_n)1_{T=n}$$
$$\leq M_n$$

with equality if equality holds in (4.2). We note that M_n is not necessarily integrable. Nevertheless, it still follows that

$$\psi_i = \mathbb{E}_i(M_0) \geq \mathbb{E}_i(M_n) = \phi_i(n-1) + \mathbb{E}_i\big(\psi(X_n)1_{T\geq n}\big)$$

with equality if equality holds in (4.2).

For continuous-time chains there is a result analogous to Theorem 4.2.3. We have to state it slightly differently because when ϕ takes infinite values the equations (4.3) may involve subtraction of infinities, and therefore not make sense. Although the conclusion then appears to depend on the finiteness of ϕ, which is *a priori* unknown, we can still use the result to determine ϕ_i in all cases. To do this we restrict our attention to the set of states J accessible from i. If the linear equations have a finite non-negative solution on J, then $(\phi_j : j \in J)$ is the minimal such solution. If not, then $\phi_j = \infty$ for some $j \in J$, which forces $\phi_i = \infty$, since i leads to j.

Theorem 4.2.4. *Assume that $(X_t)_{t\geq 0}$ is minimal, and that $(c_i : i \in D)$ and $(f_i : i \in \partial D)$ are non-negative. Set*

$$\phi_i = \mathbb{E}_i\left(\int_0^T c(X_t)dt + f(X_T)1_{T<\infty}\right)$$

where T is the hitting time of ∂D. Then $\phi = (\phi_i : i \in I)$, if finite, is the minimal non-negative solution to

$$\begin{cases} -Q\phi = c & \text{in } D \\ \phi = f & \text{in } \partial D. \end{cases} \quad (4.3)$$

If $\phi_i = \infty$ for some i, then (4.3) has no finite non-negative solution. Moreover, if $\mathbb{P}_i(T < \infty) = 1$ for all i, then (4.3) has at most one bounded solution.

Proof. Denote by $(Y_n)_{n \geq 0}$ and $S_1, S_2, \ldots$ the jump chain and holding times of $(X_t)_{t \geq 0}$, and by Π the jump matrix. Then

$$\int_0^T c(X_t)dt + f(X_T)1_{T<\infty} = \sum_{n<N} c(Y_n)S_{n+1} + f(Y_N)1_{N<\infty}$$

where N is the first time $(Y_n)_{n \geq 0}$ hits ∂D, and where we use the convention $0 \times \infty = 0$ on the right. We have

$$\mathbb{E}\big(c(Y_n)S_{n+1} \mid Y_n = j\big) = \tilde{c}_j = \begin{cases} c_j/q_j & \text{if } c_j > 0 \\ 0 & \text{if } c_j = 0, \end{cases}$$

so, by Fubini's theorem

$$\phi_i = \mathbb{E}_i \Big(\sum_{n<N} \tilde{c}(Y_n) + f(Y_N)1_{N<\infty} \Big).$$

By Theorem 4.2.3, ϕ is therefore the minimal non-negative solution to

$$\begin{cases} \phi = \Pi\phi + \tilde{c} & \text{in } D \\ \phi = f & \text{in } \partial D, \end{cases} \quad (4.4)$$

which equations have at most one bounded solution if $\mathbb{P}_i(N < \infty) = 1$ for all i. Since the finite solutions of (4.4) are exactly the finite solutions of (4.3), and since N is finite whenever T is finite, this proves the result. □

It is natural in some economic applications to apply to future costs a discount factor $\alpha \in (0,1)$ or rate $\lambda \in (0,\infty)$, corresponding to an interest rate. *Potentials with discounted costs* may also be calculated by linear equations; indeed the discounting actually makes the analysis easier.

Theorem 4.2.5. *Suppose that $(c_i : i \in I)$ is bounded. Set*

$$\phi_i = \mathbb{E}_i \sum_{n=0}^{\infty} \alpha^n c(X_n)$$

then $\phi = (\phi_i : i \in I)$ is the unique bounded solution to

$$\phi = \alpha P \phi + c.$$

Proof. Suppose that $|c_i| \leq C$ for all i, then

$$|\phi_i| \leq C \sum_{n=0}^{\infty} \alpha^n = C/(1-\alpha)$$

so ϕ is bounded. By the Markov property

$$\mathbb{E}\left(\sum_{n=1}^{\infty} \alpha^{n-1} c(X_n) \bigg| X_1 = j\right) = \mathbb{E}_j \sum_{n=0}^{\infty} \alpha^n c(X_n) = \phi_j.$$

Then

$$\phi_i = \mathbb{E}_i \sum_{n=0}^{\infty} \alpha^n c(X_n)$$

$$= c_i + \alpha \sum_{j \in I} p_{ij} \mathbb{E}\left(\sum_{n=1}^{\infty} \alpha^{n-1} c(X_n) \bigg| X_1 = j\right)$$

$$= c_i + \alpha \sum_{j \in I} p_{ij} \phi_j,$$

so

$$\phi = c + \alpha P \phi.$$

On the other hand, suppose that ψ is bounded and also that $\psi = c + \alpha P \psi$. Set $M = \sup_i |\psi_i - \phi_i|$, then $M < \infty$. But

$$\psi - \phi = \alpha P(\psi - \phi)$$

so

$$|\psi_i - \phi_i| \leq \alpha \sum_{j \in I} p_{ij} |\psi_j - \phi_j| \leq \alpha M.$$

Hence $M \leq \alpha M$, which forces $M = 0$ and $\psi = \phi$. □

We have a similar looking result for continuous time, which however lies a little deeper, because it really corresponds to a version of the discrete-time result where the discount factor may depend on the current state.

Theorem 4.2.6. *Assume that $(X_t)_{t \geq 0}$ is non-explosive. Suppose that $(c_i : i \in I)$ is bounded. Set*

$$\phi_i = \mathbb{E}_i \int_0^{\infty} e^{-\lambda t} c(X_t) dt,$$

then $\phi = (\phi_i : i \in I)$ is the unique bounded solution to

$$(\lambda - Q)\phi = c. \tag{4.5}$$

Proof. Assume for now that c is non-negative. Introduce a new state ∂ with $c_\partial = 0$. Let T be an independent $E(\lambda)$ random variable and define

$$\widetilde{X}_t = \begin{cases} X_t & \text{for } t < T \\ \partial & \text{for } t \geq T. \end{cases}$$

Then $(\widetilde{X}_t)_{t \geq 0}$ is a Markov chain on $I \cup \{\partial\}$ with modified transition rates

$$\widetilde{q}_i = q_i + \lambda, \quad \widetilde{q}_{i\partial} = \lambda, \quad \widetilde{q}_\partial = 0.$$

Also T is the hitting time of ∂, and is finite with probability 1. By Fubini's theorem

$$\phi_i = \mathbb{E}_i \int_0^T c(\widetilde{X}_t) dt.$$

Suppose $c_i \leq C$ for all i, then

$$\phi_i \leq C \int_0^\infty e^{-\lambda t} dt \leq C/\lambda,$$

so ϕ is bounded. Hence, by Theorem 4.2.4, ϕ is the unique bounded solution to

$$-\widetilde{Q}\phi = c,$$

which is the same as (4.5).

When c takes negative values we can apply the preceding argument to the potentials

$$\phi_i^\pm = \mathbb{E}_i \int_0^\infty e^{-\lambda t} c^\pm(X_t) dt$$

where $c_i^\pm = (\pm c) \vee 0$. Then $\phi = \phi^+ - \phi^-$ so ϕ is bounded. We have

$$(\lambda - Q)\phi^\pm = c^\pm$$

so, subtracting

$$(\lambda - Q)\phi = c.$$

Finally, if ψ is bounded and $(\lambda - Q)\psi = c$, then $(\lambda - Q)(\psi - \phi) = 0$, so $\psi - \phi$ is the unique bounded solution for the case when $c = 0$, which is 0. □

The point of view underlying the last four theorems was that we were interested in a given potential associated to a Markov chain, and wished to calculate it. We shall now take a brief look at some structural aspects of the set of all potentials of a given Markov chain. What we describe is just the simplest case of a structure of great generality. First we shall look at the Green matrix, and then at the role of the boundary.

4.2 Potential theory

Let us consider potentials with non-negative costs c, and without boundary. The potential is defined by

$$\phi_i = \mathbb{E}_i \sum_{n=0}^{\infty} c(X_n)$$

in discrete time, and in continuous time

$$\phi_i = \mathbb{E}_i \int_0^{\infty} c(X_t)dt.$$

By Fubini's theorem we have

$$\phi_i = \sum_{n=0}^{\infty} \mathbb{E}_i c(X_n) = \sum_{n=0}^{\infty} (P^n c)_i = (Gc)_i$$

where $G = (g_{ij} : i, j \in I)$ is the *Green* matrix

$$G = \sum_{n=0}^{\infty} P^n.$$

Similarly, in continuous time $\phi = Gc$, with

$$G = \int_0^{\infty} P(t)dt.$$

Thus, once we know the Green matrix, we have explicit expressions for all potentials of the Markov chain. The Green matrix is also called the *fundamental solution* of the linear equations (4.1) and (4.3). The jth column $(g_{ij} : i \in I)$ is itself a potential. We have

$$g_{ij} = \mathbb{E}_i \sum_{n=0}^{\infty} 1_{X_n = j}$$

in discrete time, and in continuous time

$$g_{ij} = \mathbb{E}_i \int_0^{\infty} 1_{X_t = j} dt.$$

Thus g_{ij} is the expected total time in j starting from i. These quantities have already appeared in our discussions of transience and recurrence in Sections 1.5 and 2.11: we know that $g_{ij} = \infty$ if and only if i leads to j and j is recurrent. Indeed, in discrete time

$$g_{ij} = h_i^j / (1 - f_j)$$

where h_i^j is the probability of hitting j from i, and f_j is the return probability for j. The formula for continuous time is

$$g_{ij} = h_i^j/q_j(1-f_j).$$

For potentials with discounted costs the situation is similar: in discrete time

$$\phi_i = \mathbb{E}_i \sum_{n=0}^{\infty} \alpha^n c(X_n) = \sum_{n=0}^{\infty} \alpha^n \mathbb{E}_i c(X_n) = (R_\alpha c)_i$$

where

$$R_\alpha = \sum_{n=0}^{\infty} \alpha^n P^n,$$

and in continuous time

$$\phi_i = \mathbb{E}_i \int_0^\infty e^{-\lambda t} c(X_t) dt = \int_0^\infty e^{-\lambda t} \mathbb{E}_i c(X_t) dt = (R_\lambda c)_i$$

where

$$R_\lambda = \int_0^\infty e^{-\lambda t} P(t) dt.$$

We call $(R_\alpha : \alpha \in (0,1))$ and $(R_\lambda : \lambda \in (0,\infty))$ the *resolvent* of the Markov chain. Unlike the Green matrix the resolvent is always finite. Indeed, for finite state-space we have

$$R_\alpha = (I - \alpha P)^{-1}$$

and

$$R_\lambda = (\lambda I - Q)^{-1}.$$

We return to the general case, with boundary ∂D. Any bounded function $(\phi_i : i \in I)$ for which

$$\phi = P\phi \quad \text{in } D$$

is called *harmonic* in D. Our object now is to examine the relation between non-negative functions, harmonic in D, and the boundary ∂D. Here are two examples.

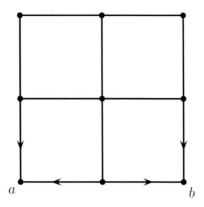

Example 4.2.7

Consider the random walk $(X_n)_{n\geq 0}$ on the above graph, where each allowable transition is made with equal probability. States a and b are absorbing. We set $\partial D = \{a, b\}$. Let h_i^a denote the absorption probability for a, starting from i. By the method of Section 1.3 we find

$$h^a = \begin{pmatrix} 3/5 & 1/2 & 2/5 \\ 7/10 & 1/2 & 3/10 \\ 1 & 1/2 & 0 \end{pmatrix}$$

where we have written the vector h^a as a matrix, corresponding in an obvious way to the state-space. The linear equations for the vector h^a read

$$\begin{cases} h^a = Ph^a & \text{in } D \\ h_a^a = 1, \; h_b^a = 0. \end{cases}$$

Thus we can find two non-negative functions h^a and h^b, harmonic in D, but with different boundary values. In fact, the most general non-negative harmonic function ϕ in D satisfies

$$\begin{cases} \phi = P\phi & \text{in } D \\ \phi = f & \text{in } \partial D \end{cases}$$

where $f_a, f_b \geq 0$, and this implies

$$\phi = f_a h^a + f_b h^b.$$

Thus the boundary points a and b give us extremal generators h^a and h^b of the set of all non-negative harmonic functions.

Example 4.2.8

Consider the random walk $(X_n)_{n\geq 0}$ on $\mathbb{Z}$ which jumps towards 0 with probability q and jumps away from 0 with probability $p = 1 - q$, except that at 0 it jumps to -1 or 1 with probability $1/2$. We choose $p > q$ so that the walk is transient. In fact, starting from 0, we can show that $(X_n)_{n\geq 0}$ is equally likely to end up drifting to the left or to the right, at speed $p - q$. Let us consider the problem of determining for $(X_n)_{n\geq 0}$ the set C of all non-negative harmonic functions ϕ. We must have

$$\phi_i = p\phi_{i+1} + q\phi_{i-1} \quad \text{for } i = 1, 2, \ldots,$$
$$\phi_0 = \tfrac{1}{2}\phi_1 + \tfrac{1}{2}\phi_{-1},$$
$$\phi_i = q\phi_{i+1} + p\phi_{i-1} \quad \text{for } i = -1, -2, \ldots.$$

The first equation has general solution

$$\phi_i = A + B(1 - (q/p)^i) \quad \text{for } i = 0, 1, 2, \ldots,$$

which is non-negative provided $A + B \geq 0$. Similarly, the third equation has general solution

$$\phi_i = A' + B'(1 - (q/p)^{-i}) \quad \text{for } i = 0, -1, -2, \ldots,$$

non-negative provided $A' + B' \geq 0$. To obtain a general harmonic function we must match the values ϕ_0 and satisfy $\phi_0 = (\phi_1 + \phi_{-1})/2$. This forces $A = A'$ and $B + B' = 0$. It follows that all non-negative harmonic functions have the form

$$\phi = f^- h^- + f^+ h^+$$

where $f^-, f^+ \geq 0$ and where $h_i^- = h_{-i}^+$ and

$$h_i^+ = \begin{cases} \tfrac{1}{2} + \tfrac{1}{2}(1 - (q/p)^i) & \text{for } i = 0, 1, 2, \ldots, \\ \tfrac{1}{2} - \tfrac{1}{2}(1 - (q/p)^{-i}) & \text{for } i = -1, -2, \ldots. \end{cases}$$

In the preceding example the generators of C were in one-to-one correspondence with the points of the boundary – the possible places for the chain to end up. In this example there is no boundary, *but the generators of C still correspond to the two possibilities for the long-time behaviour of the chain.* For we have

$$h_i^+ = \mathbb{P}_i(X_n \to \infty \text{ as } n \to \infty).$$

The suggestion of this example, which is fully developed in other works, is that the set of non-negative harmonic functions may be used to identify a generalized notion of boundary for Markov chains, which sometimes just consists of points in the state-space, but more generally corresponds to the varieties of possible limiting behaviour for X_n as $n \to \infty$. See, for example, *Markov Chains* by D. Revuz (North-Holland, Amsterdam, 1984).

We cannot begin to give the general theory corresponding to Example 4.2.8, but we can draw some general conclusions from Theorem 4.2.3 when the situation is more like Example 4.2.7. Suppose we have a Markov chain $(X_n)_{n \geq 0}$ with absorbing boundary ∂D. Set

$$h_i^\partial = \mathbb{P}_i(T < \infty)$$

where T is the hitting time of ∂D. Then by the methods of Section 1.3 we have

$$\begin{cases} h^\partial = Ph^\partial & \text{in } D \\ h^\partial = 1 & \text{in } \partial D. \end{cases} \quad (4.6)$$

Note that $h_i^\partial = 1$ for all i always gives a possible solution. Hence if (4.6) has a unique bounded solution then $h_i^\partial = \mathbb{P}_i(T < \infty) = 1$ for all i. Conversely, if $\mathbb{P}_i(T < \infty) = 1$ for all i, then, as we showed in Theorem 4.2.3, (4.6) has a unique bounded solution. Indeed, we showed more generally that this condition implies that

$$\begin{cases} \phi = P\phi + c & \text{in } D \\ \phi = f & \text{in } \partial D \end{cases}$$

has at most one bounded solution, and since

$$\phi_i = \mathbb{E}_i \left(\sum_{n<T} c(X_n) + f(X_T) 1_{T<\infty} \right) \quad (4.7)$$

is the minimal solution, any bounded solution is given by (4.7). Suppose from now on that $\mathbb{P}_i(T < \infty) = 1$ for all i. Let ϕ be a bounded non-negative function, harmonic in D, with boundary values $\phi_i = f_i$ for $i \in \partial D$. Then, by monotone convergence

$$\phi_i = \mathbb{E}_i\big(f(X_T)\big) = \sum_{j \in \partial D} f_j \mathbb{P}_i(X_T = j).$$

Hence every bounded harmonic function is determined by its boundary values and, indeed

$$\phi = \sum_{j \in \partial D} f_j h^j,$$

where

$$h_i^j = \mathbb{P}_i(X_T = j).$$

Just as in Example 4.2.7, the hitting probabilities for boundary states form a set of extremal generators for the set of all bounded non-negative harmonic functions.

Exercises

4.2.1 Consider a discrete-time Markov chain $(X_n)_{n\geq 0}$ and the potential ϕ with costs $(c_i : i \in D)$ and boundary values $(f_i : i \in \partial D)$. Set

$$\widetilde{X}_n = \begin{cases} X_n & \text{if } n \leq T \\ \partial & \text{if } n > T, \end{cases}$$

where T is the hitting time of ∂D and ∂ is a new state. Show that $(\widetilde{X}_n)_{n\geq 0}$ is a Markov chain and determine its transition matrix.

Check that

$$\phi_i = \mathbb{E}_i \sum_{n < \widetilde{T}} c(\widetilde{X}_n) = \mathbb{E}_i \sum_{n=0}^{\infty} c(\widetilde{X}_n)$$

where $\widetilde{T} = T + 1$ and where we set $c_i = f_i$ on ∂D and $c_\partial = 0$. This shows that a general potential may always be considered as a potential with boundary value zero or, indeed, without boundary at all.

Can you find a similar reduction for continuous-time chains?

4.2.2 Prove the fact claimed in Example 4.2.8 that

$$h_i^+ = \mathbb{P}_i(X_n \to \infty \text{ as } n \to \infty).$$

4.2.3 Let $(c_i : i \in I)$ be a non-negative function. Partition I as $D \cup \partial D$ and suppose that the linear equations

$$\begin{cases} \phi = P\phi + c & \text{in } D \\ \phi = 0 & \text{in } \partial D \end{cases}$$

have a unique bounded solution. Show that the Markov chain $(X_n)_{n\geq 0}$ with transition matrix P is certain to hit ∂D.

Consider now a new partition $\widetilde{D} \cup \partial\widetilde{D}$, where $\widetilde{D} \subseteq D$. Show that the linear equations

$$\begin{cases} \psi = P\psi + c & \text{in } \widetilde{D} \\ \psi = 0 & \text{in } \partial\widetilde{D} \end{cases}$$

also have a unique bounded solution, and that

$$\psi_i \leq \phi_i \quad \text{for all } i \in I.$$

4.3 Electrical networks

An electrical network has a countable set I of *nodes*, each node i having a *capacity* $\pi_i > 0$. Some nodes are joined by *wires*, the wire between i and j having *conductivity* $a_{ij} = a_{ji} \geq 0$. Where no wire joins i to j we take $a_{ij} = 0$. In practice, each 'wire' contains a resistor, which determines the conductivity as the reciprocal of its resistance. Each node i holds a certain *charge* χ_i, which determines its *potential* ϕ_i by

$$\chi_i = \phi_i \pi_i.$$

A *current* or *flow of charge* is any matrix $(\gamma_{ij} : i, j \in I)$ with $\gamma_{ij} = -\gamma_{ji}$. Physically it is found that the current γ_{ij} from i to j obeys *Ohm's law:*

$$\gamma_{ij} = a_{ij}(\phi_i - \phi_j).$$

Thus charge flows from nodes of high potential to nodes of low potential.

The first problem in electrical networks is to determine equilibrium flows and potentials, subject to given external conditions. The nodes are partitioned into two sets D and ∂D. External connections are made at the nodes in ∂D and possibly at some of the nodes in D. These have the effect that each node $i \in \partial D$ is held at a given potential f_i, and that a given current g_i enters the network at each node $i \in D$. The case where $g_i = 0$ corresponds to a node with no external connection. In equilibrium, current may also enter or leave the network through ∂D, but here it is not the current but the potential which is determined externally.

Given a flow $(\gamma_{ij} : i, j \in I)$ we shall write γ_i for the *total flow from i to the network*:

$$\gamma_i = \sum_{j \in I} \gamma_{ij}.$$

In equilibrium the charge at each node is constant, so

$$\gamma_i = g_i \quad \text{for } i \in D.$$

Therefore, by Ohm's law, any equilibrium potential $\phi = (\phi_i : i \in I)$ must satisfy

$$\begin{cases} \sum_{j \in I} a_{ij}(\phi_i - \phi_j) = g_i, & \text{for } i \in D \\ \phi_i = f_i, & \text{for } i \in \partial D. \end{cases} \quad (4.8)$$

There is a simple correspondence between electrical networks and reversible Markov chains in continuous-time, given by

$$a_{ij} = \pi_i q_{ij} \quad \text{for } i \neq j.$$

We shall assume that the total conductivity at each node is finite:

$$a_i = \sum_{j \neq i} a_{ij} < \infty.$$

Then $a_i = \pi_i q_i = -\pi_i q_{ii}$. The capacities π_i are the components of an invariant measure, and the symmetry of a_{ij} corresponds to the detailed balance equations. The equations for an equilibrium potential may now be written in a form familiar from the preceding section:

$$\begin{cases} -Q\phi = c & \text{in } D \\ \phi = f & \text{in } \partial D, \end{cases} \quad (4.9)$$

where $c_i = g_i/\pi_i$. It is natural that c appears here and not g, because ct and f have the same physical dimensions. We know that these equations may fail to have a unique solution, indicating the interesting possibility that there may be more than one equilibrium potential. However, to keep matters simple here, *we shall assume that I is finite, that the network is connected, and that ∂D is non-empty.* This is enough to ensure uniqueness of potentials. Then, by Theorem 4.2.4, the equilibrium potential is given by

$$\phi_i = \mathbb{E}_i \left(\int_0^T c(X_t) dt + f(X_T) \right) \quad (4.10)$$

where T is the hitting time of ∂D.

In fact, the case where ∂D is empty may be dealt with as follows: we must have

$$\sum_{i \in I} g_i = 0$$

or there is no possibility of equilibrium; pick one node k, set $\partial D = \{k\}$, and replace the condition $\gamma_k = g_k$ by $\phi_i = 0$. The new problem is equivalent to the old, but now ∂D is non-empty.

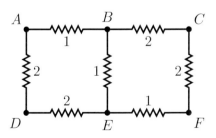

Example 4.3.1

Determine the equilibrium current in the network shown on the preceding page when unit current enters at A and leaves at F. The conductivities are shown on the diagram. Let us set $\phi_A = 1$ and $\phi_F = 0$. This will result in some flow from A to F, which we can scale to get a unit flow. By symmetry, $\phi_E = 1 - \phi_B$ and $\phi_D = 1 - \phi_C$. Then, by Ohm's law, since the total current leaving B and C must vanish

$$(\phi_B - \phi_A) + (\phi_B - \phi_E) + 2(\phi_B - \phi_C) = 0,$$
$$2(\phi_C - \phi_F) + 2(\phi_C - \phi_B) = 0.$$

Hence, $\phi_B = 1/2$ and $\phi_C = 1/4$, and the associated flow is given by $\gamma_{AB} = 1/2, \gamma_{BC} = 1/2, \gamma_{CF} = 1/2, \gamma_{BE} = 0$. In fact, we were lucky – no scaling was necessary.

Note that the node capacities do not affect the problem we considered. Let us arbitrarily assign to each node a capacity 1. Then there is an associated Markov chain and, according to (4.10), the equilibrium potential is given by

$$\phi_i = \mathbb{E}_i(1_{X_T = A}) = \mathbb{P}_i(X_T = A)$$

where T is the hitting time of $\{A, F\}$. Different node capacities result in different Markov chains, but the same jump chain and hence the same hitting probabilities.

Here is a general result expressing equilibrium potentials, flows and charges in terms of the associated Markov chain.

Theorem 4.3.2. *Consider a finite network with external connections at two nodes A and B, and the associated Markov chain $(X_t)_{t \geq 0}$.*
(a) The unique equilibrium potential ϕ with $\phi_A = 1$ and $\phi_B = 0$ is given by

$$\phi_i = \mathbb{P}_i(T_A < T_B)$$

where T_A and T_B are the hitting times of A and B.
(b) The unique equilibrium flow γ with $\gamma_A = 1$ and $\gamma_B = -1$ is given by

$$\gamma_{ij} = \mathbb{E}_A(\Gamma_{ij} - \Gamma_{ji})$$

where Γ_{ij} is the number of times that $(X_t)_{t \geq 0}$ jumps from i to j before hitting B.
(c) The charge χ associated with γ, subject to $\chi_B = 0$, is given by

$$\chi_i = \mathbb{E}_A \int_0^{T_B} 1_{\{X_t = i\}} dt.$$

Proof. The formula for ϕ is a special case of (4.10), where $c = 0$ and $f = 1_{\{A\}}$. We shall prove (b) and (c) together. Observe that if $X_0 = A$ then

$$\sum_{j \neq i}(\Gamma_{ij} - \Gamma_{ji}) = \begin{cases} 1 & \text{if } i = A \\ 0 & \text{if } i \notin \{A, B\} \\ -1 & \text{if } i = B \end{cases}$$

so if $\gamma_{ij} = \mathbb{E}_A(\Gamma_{ij} - \Gamma_{ji})$ then γ is a unit flow from A to B. We have

$$\Gamma_{ij} = \sum_{n=0}^{\infty} 1_{\{Y_n = i, Y_{n+1} = j, n < N_B\}}$$

where N_B is the hitting time of B for the jump chain $(Y_n)_{n \geq 0}$. So, by the Markov property of the jump chain

$$\mathbb{E}_A(\Gamma_{ij}) = \sum_{n=0}^{\infty} \mathbb{P}_A(Y_n = i, Y_{n+1} = j, n < N_B)$$

$$= \sum_{n=0}^{\infty} \mathbb{P}_A(Y_n = i, n < N_B)\pi_{ij}.$$

Set

$$\chi_i = \mathbb{E}_A \int_0^{T_B} 1_{\{X_t = i\}} dt$$

and consider the associated potential $\psi_i = \chi_i/\pi_i$. Then

$$\chi_i q_{ij} = \chi_i q_i \pi_{ij} = \sum_{n=0}^{\infty} \mathbb{P}_A(Y_n = i, n < N_B)\pi_{ij} = \mathbb{E}_A(\Gamma_{ij})$$

so

$$(\psi_i - \psi_j)a_{ij} = \chi_i q_{ij} - \chi_j q_{ij} = \gamma_{ij}.$$

Hence $\psi = \phi$, γ is the equilibrium unit flow and χ the associated charge, as required. $\square$

The interpretation of potential theory in terms of electrical networks makes it natural to consider notions of *energy*. We define for a potential $\phi = (\phi_i : i \in I)$ and a flow $\gamma = (\gamma_{ij} : i, j \in I)$

$$E(\phi) = \tfrac{1}{2} \sum_{i,j \in I} (\phi_i - \phi_j)^2 a_{ij},$$

$$I(\gamma) = \tfrac{1}{2} \sum_{i,j \in I} \gamma_{ij}^2 a_{ij}^{-1}.$$

4.3 Electrical networks

The 1/2 means that each wire is counted once. When ϕ and γ are related by Ohm's law we have

$$E(\phi) = \tfrac{1}{2} \sum_{i,j \in I} (\phi_i - \phi_j)\gamma_{ij} = I(\gamma)$$

and $E(\phi)$ is found physically to give the rate of dissipation of energy, as heat, by the network. Moreover, we shall see that certain equilibrium potentials and flows determined by Ohm's law minimize these energy functions. This characteristic of energy minimization can indeed replace Ohm's law as the fundamental physical principle.

Theorem 4.3.3. *The equilibrium potential and flow may be determined as follows.*

(a) *The equilibrium potential $\phi = (\phi_i : i \in I)$ with boundary values $\phi_i = f_i$ for $i \in \partial D$ and no current sources in D is the unique solution to*

$$\begin{aligned} \text{minimize} \quad & E(\phi) \\ \text{subject to} \quad & \phi_i = f_i \quad \text{for } i \in \partial D. \end{aligned}$$

(b) *The equilibrium flow $\gamma = (\gamma_{ij} : i, j \in I)$ with current sources $\gamma_i = g_i$ for $i \in D$ and boundary potential zero is the unique solution to*

$$\begin{aligned} \text{minimize} \quad & I(\gamma) \\ \text{subject to} \quad & \gamma_i = g_i \quad \text{for } i \in D. \end{aligned}$$

Proof. For any potential $\phi = (\phi_i : i \in I)$ and any flow $\gamma = (\gamma_{ij} : i, j \in I)$ we have

$$\sum_{i,j \in I} (\phi_i - \phi_j)\gamma_{ij} = 2 \sum_{i \in I} \phi_i \gamma_i.$$

(a) Denote by $\phi = (\phi_i : i \in I)$ and by $\gamma = (\gamma_{ij} : i, j \in I)$ the equilibrium potential and flow. We have $\gamma_i = 0$ for $i \in D$. We can write any potential in the minimization problem in the form $\phi + \varepsilon$, where $\varepsilon = (\varepsilon_i : i \in I)$ with $\varepsilon_i = 0$ for $i \in \partial D$. Then

$$\sum_{i,j \in I} (\varepsilon_i - \varepsilon_j)(\phi_i - \phi_j)a_{ij} = \sum_{i,j \in I} (\varepsilon_i - \varepsilon_j)\gamma_{ij} = 2 \sum_{i \in I} \varepsilon_i \gamma_i = 0$$

so

$$E(\phi + \varepsilon) = E(\phi) + E(\varepsilon) \geq E(\phi)$$

with equality only if $\varepsilon = 0$.

(b) Denote by $\phi = (\phi_i : i \in I)$ and by $\gamma = (\gamma_{ij} : i,j \in I)$ the equilibrium potential and flow. We have $\phi_i = 0$ for $i \in \partial D$. We can write any flow in the minimization problem in the form $\gamma + \delta$, where $\delta = (\delta_{ij} : i,j \in I)$ is a flow with $\delta_i = 0$ for $i \in D$. Then

$$\sum_{i,j \in I} \gamma_{ij} \delta_{ij} a_{ij}^{-1} = \sum_{i,j \in I} (\phi_i - \phi_j) \delta_{ij} = 2 \sum_{i \in I} \phi_i \delta_i = 0,$$

so

$$I(\gamma + \delta) = I(\gamma) + I(\delta) \geq I(\delta)$$

with equality only if $\delta = 0$. □

The following reformulation of part (a) of the preceding result states that harmonic functions minimize energy.

Corollary 4.3.4. *Suppose that* $\phi = (\phi_i : i \in I)$ *satisfies*

$$\begin{cases} Q\phi = 0 & \text{in } D \\ \phi = f & \text{in } \partial D. \end{cases}$$

Then ϕ is the unique solution to

$$\text{minimize} \quad E(\phi)$$
$$\text{subject to} \quad \phi = f \quad \text{in } \partial D.$$

An important feature of electrical networks is that networks with a small number of external connections look like networks with a small number of nodes altogether. In fact, given any network, there is always another network of wires joining the externally connected nodes alone, equivalent in its response to external flows and potentials.

Let $J \subseteq I$. We say that $\bar{a} = (\bar{a}_{ij} : i,j \in J)$ is an *effective conductivity* on J if, for all potentials $f = (f_i : i \in J)$, the external currents into J when J is held at potential f are the same for $(J, \bar{a})$ as for (I, a). We know that f determines an equilibrium potential $\phi = (\phi_i : i \in I)$ by

$$\begin{cases} \sum_{j \in I} (\phi_i - \phi_j) a_{ij} = 0 & \text{for } i \notin J \\ \phi_i = f_i & \text{for } i \in J. \end{cases}$$

Then $\bar{a}$ is an effective conductivity if, for all f, for $i \in J$ we have

$$\sum_{j \in I} (\phi_i - \phi_j) a_{ij} = \sum_{j \in J} (f_i - f_j) \, \bar{a}_{ij}.$$

For a conductivity matrix $\bar{a}$ on J, for a potential $f = (f_i : i \in J)$ and a flow $\delta = (\delta_{ij} : i,j \in J)$ we set

$$\overline{E}(f) = \tfrac{1}{2} \sum_{i,j \in J} (f_i - f_j)^2 \bar{a}_{ij},$$

$$\overline{I}(\delta) = \tfrac{1}{2} \sum_{i,j \in J} \delta_{ij}^2 \bar{a}_{ij}^{-1}.$$

4.3 Electrical networks

Theorem 4.3.5. *There is a unique effective conductivity $\overline{a}$ given by*

$$\overline{a}_{ij} = a_{ij} + \sum_{k \notin J} a_{ik} \phi_k^j$$

where for each $j \in J$, $\phi^j = (\phi_i^j : i \in I)$ is the potential defined by

$$\begin{cases} \sum_{k \in I} (\phi_i^j - \phi_k^j) a_{ik} = 0 & \text{for } i \notin J \\ \phi_i^j = \delta_{ij} & \text{for } i \in J. \end{cases} \quad (4.11)$$

Moreover, $\overline{a}$ is characterized by the Dirichlet variational principle

$$\overline{E}(f) = \inf_{\phi_i = f_i \text{ on } J} E(\phi),$$

and also by the Thompson variational principle

$$\inf_{\delta_i = g_i \text{ on } J} \overline{I}(\delta) = \inf_{\gamma_i = \begin{cases} g_i \text{ on } J \\ 0 \text{ off } J \end{cases}} I(\gamma).$$

Proof. Given $f = (f_i : i \in J)$, define $\phi = (\phi_i : i \in I)$ by

$$\phi_i = \sum_{j \in J} f_j \phi_i^j$$

then ϕ is the equilibrium potential given by

$$\begin{cases} \sum_{j \in I} a_{ij}(\phi_i - \phi_j) = 0 & \text{for } i \notin J \\ \phi_i = f_i & \text{for } i \in J, \end{cases}$$

and, by Corollary 4.3.4, ϕ solves

$$\begin{aligned} \text{minimize} \quad & E(\phi) \\ \text{subject to} \quad & \phi_i = f_i \quad \text{for } i \in J. \end{aligned}$$

We have, for $i \in J$

$$\sum_{j \in I} a_{ij} \phi_j = \sum_{j \in J} a_{ij} f_j + \sum_{k \notin J} \sum_{j \in J} a_{ik} \phi_k^j f_j = \sum_{j \in J} \overline{a}_{ij} f_j.$$

In particular, taking $f \equiv 1$ we obtain

$$\sum_{j \in I} a_{ij} = \sum_{j \in J} \overline{a}_{ij}.$$

Hence we have equality of external currents:
$$\sum_{j\in I}(\phi_i - \phi_j)a_{ij} = \sum_{j\in J}(f_i - f_j)\overline{a}_{ij}.$$

Moreover, we also have equality of energies:
$$\sum_{i,j\in I}(\phi_i - \phi_j)^2 a_{ij} = 2\sum_{i\in I}\phi_i\sum_{j\in I}(\phi_i - \phi_j)a_{ij}$$
$$= 2\sum_{i\in J}f_i\sum_{j\in J}(f_i - f_j)\overline{a}_{ij} = \sum_{i,j\in J}(f_i - f_j)^2 \overline{a}_{ij}.$$

Finally, if $g_{ij} = (f_i - f_j)\overline{a}_{ij}$ and $\gamma_{ij} = (\phi_i - \phi_j)a_{ij}$, then
$$\sum_{i,j\in I}\gamma_{ij}^2 a_{ij}^{-1} = \sum_{i,j\in I}(\phi_i - \phi_j)^2 a_{ij}$$
$$= \sum_{i,j\in J}(f_i - f_j)^2 \overline{a}_{ij} = \sum_{i,j\in J}g_{ij}^2 \overline{a}_{ij}^{-1},$$

so, by Theorem 4.3.3, for any flow $\delta = (\delta_{ij} : i,j \in I)$ with $\delta_i = g_i$ for $i \in J$ and $\delta_i = 0$ for $i \notin J$, we have
$$\sum_{i,j\in I}\delta_{ij}^2 a_{ij}^{-1} \geq \sum_{i,j\in J}g_{ij}^2 \overline{a}_{ij}^{-1}. \qquad \square$$

Effective conductivity is also related to the associated Markov chain $(X_t)_{t\geq 0}$ in an interesting way. Define the *time spent in J*
$$A_t = \int_0^t \mathbf{1}_{\{X_s \in J\}} ds$$

and a time-changed process $(\overline{X}_t)_{t\geq 0}$ by
$$\overline{X}_t = X_{\tau(t)}$$

where
$$\tau(t) = \inf\{s \geq 0 : A_s > t\}.$$

We obtain $(\overline{X}_t)_{t\geq 0}$ by observing $(X_t)_{t\geq 0}$ whilst in J, and stopping the clock whilst $(X_t)_{t\geq 0}$ makes excursions outside J. This is really a transformation of the jump chain. By applying the strong Markov property to the jump chain we find that $(\overline{X}_t)_{t\geq 0}$ is itself a Markov chain, with jump matrix $\overline{\Pi}$ given by
$$\overline{\pi}_{ij} = \pi_{ij} + \sum_{k\notin J}\pi_{ik}\phi_k^j \qquad \text{for } i,j \in J,$$

where
$$\phi_k^j = \mathbb{P}_k(X_T = j)$$
and T denotes the hitting time of J. See Example 1.4.4. Hence $(\overline{X}_t)_{t\geq 0}$ has Q-matrix given by
$$\overline{q}_{ij} = q_{ij} + \sum_{k \notin J} q_{ik}\phi_k^j.$$

Since $\phi^j = (\phi_k^j : k \in I)$ is the unique solution to (4.11), this shows that
$$\pi_i \overline{q}_{ij} = \overline{a}_{ij}$$
so $(\overline{X}_t)_{t\geq 0}$ is the Markov chain on J associated with the effective conductivity $\overline{a}$.

There is much more that one can say, for example in tying up the non-equilibrium behaviour of Markov chains and electrical networks. Moreover, methods coming from one theory one provide insights into the other. For an entertaining and illuminating account of the subject, you should see *Random Walks and Electrical Networks* by P. G. Doyle and J. L. Snell (Carus Mathematical Monographs 22, Mathematical Association of America, 1984).

4.4 Brownian motion

Imagine a symmetric random walk in Euclidean space which takes infinitesimal jumps with infinite frequency and you will have some idea of Brownian motion. It is named after a botanist who observed such a motion when looking at pollen grains under a microscope. The mathematical object now called Brownian motion was actually discovered by Wiener, and is also called the Wiener process.

A discrete approximation to Euclidean space $\mathbb{R}^d$ is provided by
$$c^{-1/2}\mathbb{Z}^d = \{c^{-1/2}z : z \in \mathbb{Z}^d\}$$
where c is a large positive number. The simple symmetric random walk $(X_n)_{n\geq 0}$ on $\mathbb{Z}^d$ is a Markov chain which is by now quite familiar. We shall show that the scaled-down and speeded-up process
$$X_t^{(c)} = c^{-1/2}X_{ct}$$
is a good approximation to Brownian motion. This provides an elementary way of thinking about Brownian motion. Also, it makes it reasonable to

suppose that some properties of the random walk carry over to Brownian motion. At the end of this section we state some results which confirm that this is true to a remarkable extent.

Why is space rescaled by the square-root of the time-scaling? Well, if we hope that $X_t^{(c)}$ converges in some sense as $c \to \infty$ to a non-degenerate limit, we will at least want $\mathbb{E}[|X_t^{(c)}|^2]$ to converge to a non-degenerate limit. For $ct \in \mathbb{Z}^+$ we have

$$\mathbb{E}[|X_{ct}|^2] = ct\mathbb{E}[|X_1|^2]$$

so the square-root scaling gives

$$\mathbb{E}[|X_t^{(c)}|^2] = t\mathbb{E}[|X_1|^2]$$

which is independent of c.

We begin by defining Brownian motion, and then show that this is not an empty definition; that is to say, Brownian motions exist.

A real-valued random variable is said to have *Gaussian distribution with mean 0 and variance t* if it has density function

$$\phi_t(x) = (2\pi t)^{-1/2} \exp\{-x^2/2t\}.$$

The fundamental role of Gaussian distributions in probability derives from the following result.

Theorem 4.4.1 (Central limit theorem). Let $X_1, X_2, \ldots$ be a sequence of independent and identically distributed real-valued random variables with mean 0 and variance $t \in (0, \infty)$. Then, for all bounded continuous functions f, as $n \to \infty$ we have

$$\mathbb{E}\big[f\big((X_1 + \ldots + X_n)/\sqrt{n}\big)\big] \to \int_{\mathbb{R}} f(x)\phi_t(x)dx.$$

We shall take this result and a few other standard properties of the Gaussian distribution for granted in this section. There are many introductory texts on probability which give the full details.

A real-valued process $(X_t)_{t\geq 0}$ is said to be *continuous* if

$$\mathbb{P}(\{\omega : t \mapsto X_t(\omega) \text{ is continuous}\}) = 1.$$

A continuous real-valued process $(B_t)_{t\geq 0}$ is called a *Brownian motion* if $B_0 = 0$ and for all $0 = t_0 < t_1 < \ldots < t_n$ the increments

$$B_{t_1} - B_{t_0}, \ldots, B_{t_n} - B_{t_{n-1}}$$

are independent Gaussian random variables of mean 0 and variance

$$t_1 - t_0, \ldots, t_n - t_{n-1}.$$

The conditions made on $(B_t)_{t\geq 0}$ are enough to determine all the probabilities associated with the process. To put it properly, the law of Brownian motion, which is a measure on the set of continuous paths, is uniquely determined. However, it is not obvious that there is any such process. We need the following result.

Theorem 4.4.2 (Wiener's theorem). *Brownian motion exists.*

Proof. For $N = 0, 1, 2, \ldots$, denote by D_N the set of integer multiples of 2^{-N} in $[0, \infty)$, and denote by D the union of these sets. Let us say that $(B_t : t \in D_N)$ is a *Brownian motion indexed by D_N* if $B_0 = 0$ and for all $0 = t_0 < t_1 < \ldots < t_n$ in D_N the increments

$$B_{t_1} - B_{t_0}, \ldots, B_{t_n} - B_{t_{n-1}}$$

are independent Gaussian random variables of mean 0 and variance

$$t_1 - t_0, \ldots, t_n - t_{n-1}.$$

We suppose given, for each $t \in D$, an independent Gaussian random variable Y_t of mean 0 and variance 1. For $t \in D_0 = \mathbb{Z}^+$ set

$$B_t = Y_1 + Y_2 + \ldots + Y_t$$

then $(B_t : t \in D_0)$ is a Brownian motion indexed by D_0. We shall show how to extend this process successively to Brownian motions $(B_t : t \in D_N)$ indexed by D_N. Then $(B_t : t \in D)$ is a Brownian motion indexed by D. Next we shall show that $(B_t : t \in D)$ extends continuously to $t \in [0, \infty)$, and finally check that the extension is a Brownian motion.

Suppose we have constructed $(B_t : t \in D_{N-1})$, a Brownian motion indexed by D_{N-1}. For $t \in D_N \setminus D_{N-1}$ set $r = t - 2^{-N}$ and $s = t + 2^{-N}$ so that $r, s \in D_{N-1}$ and define

$$Z_t = 2^{-(N+1)/2} Y_t,$$
$$B_t = \tfrac{1}{2}(B_r + B_s) + Z_t.$$

We obtain two new increments:

$$B_t - B_r = \tfrac{1}{2}(B_s - B_r) + Z_t,$$
$$B_s - B_t = \tfrac{1}{2}(B_s - B_r) - Z_t.$$

We compute

$$\mathbb{E}[(B_t - B_r)^2] = \mathbb{E}[(B_s - B_t)^2] = \tfrac{1}{4}2^{-(N-1)} + 2^{-(N+1)} = 2^{-N},$$
$$\mathbb{E}[(B_t - B_r)(B_s - B_t)] = \tfrac{1}{4}2^{-(N-1)} - 2^{-(N+1)} = 0.$$

The two new increments, being Gaussian, are therefore independent and of the required variance. Moreover, being constructed from $B_s - B_r$ and Y_t, they are certainly independent of increments over intervals disjoint from (r,s). Hence $(B_t : t \in D_N)$ is a Brownian motion indexed by D_N, as required. Hence, by induction, we obtain a Brownian motion $(B_t : t \in D)$.

For each N denote by $(B_t^{(N)})_{t \geq 0}$ the continuous process obtained by linear interpolation from $(B_t : t \in D_N)$. Also, set $Z_t^{(N)} = B_t^{(N)} - B_t^{(N-1)}$. For $t \in D_{N-1}$ we have $Z_t^{(N)} = 0$. For $t \in D_N \setminus D_{N-1}$, by our construction we have

$$Z_t^{(N)} = B_t - \tfrac{1}{2}(B_{t-2^{-N}} + B_{t+2^{-N}}) = Z_t = 2^{-(N+1)/2}Y_t$$

with Y_t Gaussian of mean 0 and variance 1. Set

$$M_N = \sup_{t \in [0,1]} |Z_t^{(N)}|.$$

Then, since $(Z_t^{(N)})_{t \geq 0}$ interpolates linearly between its values on D_N, we obtain

$$M_N = \sup_{t \in (D_N \setminus D_{N-1}) \cap [0,1]} 2^{-(N+1)/2}|Y_t|.$$

There are 2^{N-1} points in $(D_N \setminus D_{N-1}) \cap [0,1]$. So for $\lambda > 0$ we have

$$\mathbb{P}(M_N > \lambda 2^{-(N+1)/2}) \leq 2^{N-1}\mathbb{P}(|Y_1| > \lambda).$$

For a random variable $X \geq 0$ and $p > 0$ we have the formula

$$\mathbb{E}(X^p) = \mathbb{E}\int_0^\infty 1_{\{X > \lambda\}} p\lambda^{p-1} d\lambda = \int_0^\infty p\lambda^{p-1}\mathbb{P}(X > \lambda) d\lambda.$$

Hence

$$2^{p(N+1)/2}\mathbb{E}(M_N^p) = \int_0^\infty p\lambda^{p-1}\mathbb{P}(2^{(N+1)/2}M_N > \lambda) d\lambda$$
$$\leq 2^{N-1}\int_0^\infty p\lambda^{p-1}\mathbb{P}(|Y_1| > \lambda) d\lambda = 2^{N-1}\mathbb{E}(|Y_1|^p)$$

and hence, for any $p > 2$

$$\mathbb{E} \sum_{N=0}^{\infty} M_n = \sum_{N=0}^{\infty} \mathbb{E}(M_N)$$

$$\leq \sum_{N=0}^{\infty} \mathbb{E}(M_N^p)^{1/p} \leq \mathbb{E}(|Y_1|^p)^{1/p} \sum_{N=0}^{\infty} (2^{(p-2)/2p})^{-N} < \infty.$$

It follows that, with probability 1, as $N \to \infty$

$$B_t^{(N)} = B_t^{(0)} + Z_t^{(1)} + \ldots + Z_t^{(N)}$$

converges uniformly in $t \in [0,1]$, and by a similar argument uniformly for t in any bounded interval. Now $B_t^{(N)}$ eventually equals B_t for any $t \in D$ and the uniform limit of continuous functions is continuous. Therefore, $(B_t : t \in D)$ has a continuous extension $(B_t)_{t \geq 0}$, as claimed.

It remains to show that the increments of $(B_t)_{t \geq 0}$ have the required joint distribution. But given $0 < t_1 < \ldots < t_n$ we can find sequences $(t_k^m)_{m \in \mathbb{N}}$ in D such that $0 < t_1^m < \ldots < t_n^m$ for all m and $t_k^m \to t_k$ for all k. Set $t_0 = t_0^m = 0$. We know that the increments

$$B_{t_1^m} - B_{t_0^m}, \ldots, B_{t_n^m} - B_{t_{n-1}^m}$$

are Gaussian of mean 0 and variance

$$t_1^m - t_0^m, \ldots, t_n^m - t_{n-1}^m.$$

Hence, using continuity of $(B_t)_{t \geq 0}$ we can let $m \to \infty$ to obtain the desired distribution for the increments

$$B_{t_1} - B_{t_0}, \ldots, B_{t_n} - B_{t_{n-1}}. \qquad \square$$

Having shown that Brownian motion exists, we now want to show how it appears as a universal scaling limit of random walks, very much as the Gaussian distribution does for sums of independent random variables.

Theorem 4.4.3. Let $(X_n)_{n \geq 0}$ be a discrete-time real-valued random walk with steps of mean 0 and variance $\sigma^2 \in (0, \infty)$. For $c > 0$ consider the rescaled process

$$X_t^{(c)} = c^{-1/2} X_{ct}$$

where the value of X_{ct} when ct is not an integer is found by linear interpolation. Then, for all m, for all bounded continuous functions $f : \mathbb{R}^m \to \mathbb{R}$ and all $0 \leq t_1 < \ldots < t_m$, we have

$$\mathbb{E}[f(X_{t_1}^{(c)}, \ldots, X_{t_m}^{(c)})] \to \mathbb{E}[f(\sigma B_{t_1}, \ldots, \sigma B_{t_m})]$$

as $c \to \infty$, where $(B_t)_{t \geq 0}$ is a Brownian motion.

Proof. The claim is that $(X^{(c)}_{t_1}, \ldots, X^{(c)}_{t_m})$ converges weakly to $(\sigma B_{t_1}, \ldots, \sigma B_{t_m})$ as $c \to \infty$. In the proof we shall take for granted some basic properties of weak convergence. First define

$$\widetilde{X}^{(c)}_t = c^{-1/2} X_{[ct]}$$

where $[ct]$ denotes the integer part of ct. Then

$$|(X^{(c)}_{t_1}, \ldots, X^{(c)}_{t_m}) - (\widetilde{X}^{(c)}_{t_1}, \ldots, \widetilde{X}^{(c)}_{t_m})| \leq c^{-1/2} |(Y_{[ct_1]+1}, \ldots, Y_{[ct_n]+1})|$$

where Y_n denotes the nth step of $(X_n)_{n \geq 0}$. The right side converges weakly to 0, so it suffices to prove the claim with $\widetilde{X}^{(c)}_t$ replacing $X^{(c)}_t$.

Consider now the increments

$$U^{(c)}_k = \widetilde{X}^{(c)}_{t_k} - \widetilde{X}^{(c)}_{t_{k-1}}, \quad Z_k = \sigma(B_{t_k} - B_{t_{k-1}})$$

for $k = 1, \ldots, m$. Since $\widetilde{X}^{(c)}_0 = B_0 = 0$ it suffices to show that $(U^{(c)}_1, \ldots, U^{(c)}_m)$ converges weakly to $(Z_1, \ldots, Z_m)$. Then since both sets of increments are independent, it suffices to show that $U^{(c)}_k$ converges weakly to Z_k for each k. But

$$U^{(c)}_k = c^{-1/2} \sum_{n=[ct_{k-1}]+1}^{[ct_k]} Y_n \sim \big(c^{-1/2} N_k(c)^{1/2}\big) N_k(c)^{-1/2} (Y_1 + \ldots + Y_{N(c)})$$

where $\sim$ denotes identity of distribution and $N_k(c) = [ct_k] - [ct_{k-1}]$. By the central limit theorem $N_k(c)^{-1/2}(Y_1 + \ldots + Y_{N(c)})$ converges weakly to $(t_k - t_{k-1})^{-1/2} Z_k$, and $\big(c^{-1/2} N_k(c)^{1/2}\big) \to (t_k - t_{k-1})^{1/2}$. Hence $U^{(c)}_k$ converges weakly to Z_k, as required. □

To summarize the last two results, we have shown, using special properties of the Gaussian distribution, that there is a continuous process $(B_t)_{t \geq 0}$ with stationary independent increments and such that B_t is Gaussian of mean 0 and variance t, for each $t \geq 0$. That was Wiener's theorem. Then, using the central limit theorem applied to the increments of a rescaled random walk, we established a sort of convergence to Brownian motion. There now follows a series of related remarks.

Note the similarity to the definition of a *Poisson process* as a right-continuous integer-valued process $(X_t)_{t \geq 0}$ starting from 0, having stationary independent increments and such that X_t is Poisson of parameter λt for each $t \geq 0$.

Given d independent Brownian motions $(B^1_t)_{t \geq 0}, \ldots, (B^d_t)_{t \geq 0}$, let us consider the $\mathbb{R}^d$-valued process $B_t = (B^1_t, \ldots, B^d_t)$. We call $(B_t)_{t \geq 0}$ a *Brownian* motion in $\mathbb{R}^d$. There is a multidimensional version of the central limit theorem which leads to a multidimensional version of Theorem 4.4.3, with no essential change in the proof. Thus if $(X_n)_{n\geq 0}$ is a random walk in $\mathbb{R}^d$ with steps of mean 0 and covariance matrix

$$V = \mathbb{E}(X_1 X_1^T)$$

and if V is finite, then for all bounded continuous functions $f : (\mathbb{R}^d)^m \to \mathbb{R}$, as $c \to \infty$ we have

$$\mathbb{E}[f(X^{(c)}_{t_1}, \ldots, X^{(c)}_{t_m})] \to \mathbb{E}[f(\sqrt{V} B_{t_1}, \ldots, \sqrt{V} B_{t_m})].$$

Here are two examples. We might take $(X_n)_{n\geq 0}$ to be the simple symmetric random walk in $\mathbb{Z}^3$, then $V = \frac{1}{3}I$. Alternatively, we might take the components of $(X_n)_{n\geq 0}$ to be three independent simple symmetric random walks in $\mathbb{Z}$, in which case $V = I$. Although these are different random walks, once the difference in variance is taken out, the result shows that in the scaling limit they behave asymptotically the same. More generally, given a random walk with a complicated step distribution, it is useful to know that on large scales all one needs to calculate is the variance (or covariance matrix). All other aspects of the step distribution become irrelevant as $c \to \infty$.

The scaling used in Theorem 4.4.3 suggests the following *scaling invariance property* of Brownian motion $(B_t)_{t\geq 0}$, which is also easy to check from the definition. For any $c > 0$ the process $(B^{(c)}_t)_{t\geq 0}$ defined by

$$B^{(c)}_t = c^{-1/2} B_{ct}$$

is a Brownian motion. Thus Brownian motion appears as a fixed point of the scaling transformation, which attracts all other finite variance symmetric random walks as $c \to \infty$.

The sense in which we have shown that $(X^{(c)}_t)_{t\geq 0}$ converges to Brownian motion is very weak, and one can with effort prove stronger forms of convergence. However, what we have proved is strong enough to ensure that $(X^{(c)}_t)_{t\geq 0}$ does not converge, in the same sense, to anything else.

The discussion to this point has not really been about the Markov property, but rather about processes with independent increments. To remedy this we must first define *Brownian motion starting from x*: this is simply any process $(B_t)_{t\geq 0}$ such that $B_0 = x$ and $(B_t - B_0)_{t\geq 0}$ is a Brownian motion (starting from 0). As a limit of Markov chains it is natural to look in Brownian motion for the structure of a Markov process. By analogy with continuous-time Markov chains we look for a *transition semigroup* $(P_t)_{t\geq 0}$

and a *generator* G. For any bounded measurable function $f : \mathbb{R}^d \to \mathbb{R}$ we have

$$\mathbb{E}_x[f(B_t)] = \mathbb{E}_0[f(x+B_t)] = \int_{\mathbb{R}^d} f(x+y)\phi_t(y_1)\ldots\phi_t(y_d)dy_1\ldots dy_d$$
$$= \int_{\mathbb{R}^d} p(t,x,y)f(y)dy$$

where

$$p(t,x,y) = (2\pi t)^{-d/2}\exp\{-|y-x|^2/2t\}.$$

This is the *transition density* for Brownian motion and the *transition semigroup* is given by

$$(P_t f)(x) = \int_{\mathbb{R}^d} p(t,x,y)f(y)dy = \mathbb{E}_x[f(B_t)].$$

To check the semigroup property $P_s P_t = P_{s+t}$ we note that

$$\mathbb{E}_x[f(B_{s+t})] = \mathbb{E}_x\left[f\left(B_s + (B_{s+t} - B_s)\right)\right]$$
$$= \mathbb{E}_x\left[P_t f(B_s)\right] = (P_s P_t f)(x)$$

where we first took the expectation over the independent increment $B_{s+t} - B_s$. For $t > 0$ it is easy to check that

$$\frac{\partial}{\partial t}p(t,x,y) = \tfrac{1}{2}\Delta_x p(t,x,y)$$

where

$$\Delta_x = \frac{\partial^2}{\partial x_1^2} + \ldots + \frac{\partial^2}{\partial x_d^2}.$$

Hence, if f has two bounded derivatives, we have

$$\frac{\partial}{\partial t}(P_t f)(x) = \int_{\mathbb{R}^d} \tfrac{1}{2}\Delta_x p(t,x,y)f(y)dy$$
$$= \int_{\mathbb{R}^d} \tfrac{1}{2}\Delta_y p(t,x,y)f(y)dy$$
$$= \int_{\mathbb{R}^d} p(t,x,y)(\tfrac{1}{2}\Delta f)(y)dy$$
$$= \mathbb{E}_x[(\tfrac{1}{2}\Delta f)(B_t)] \to \tfrac{1}{2}\Delta f(x)$$

as $t \downarrow 0$. This suggests, by analogy with continuous-time chains, that the generator, a term we have not defined precisely, should be given by

$$G = \tfrac{1}{2}\Delta.$$

Where formerly we considered vectors $(f_i : i \in I)$, now there are functions $f : \mathbb{R}^d \to \mathbb{R}$, required to have various degrees of local regularity, such as measurability and differentiability. Where formerly we considered matrices P_t and Q, now we have linear operators on functions: P_t is an integral operator, G is a differential operator.

We would like to explain the appearance of the Laplacian Δ by reference to the random walk approximation. Denote by $(X_n)_{n\geq 0}$ the simple symmetric random walk in $\mathbb{Z}^d$ and consider for $N = 1, 2, \ldots$ the rescaled process
$$X_t^{(N)} = N^{-1/2} X_{Nt}, \quad t = 0, 1/N, 2/N, \ldots.$$
For a bounded continuous function $f : \mathbb{R}^d \to \mathbb{R}$, set
$$(P_t^{(N)} f)(x) = \mathbb{E}_x[f(X_t^{(N)})], \quad x \in N^{-1/2}\mathbb{Z}^d.$$
The closest thing we have to a derivative in t at 0 for $(P_t^{(N)})_{t=0,1/N,2/N,\ldots}$ is
$$\begin{aligned} N(P_{1/N}^{(N)} f - f)(x) &= N\mathbb{E}_x[f(X_{1/N}^{(N)}) - f(X_0^{(N)})] \\ &= N\mathbb{E}_{N^{1/2}x}[f(N^{-1/2}X_1) - f(N^{-1/2}X_0)] \\ &= (N/2)\{f(x - N^{-1/2}) - 2f(x) + f(x + N^{-1/2})\}. \end{aligned}$$

If we assume that f has two bounded derivatives then, by Taylor's theorem, as $N \to \infty$,
$$f(x - N^{-1/2}) - 2f(x) + f(x + N^{-1/2}) = N^{-1}\big(\Delta f(x) + o(N)\big),$$
so
$$N(P_{1/N}^{(N)} f - f)(x) \to \tfrac{1}{2}\Delta f(x).$$

We finish by stating some results about Brownian motion which emphasise how much of the structure of Markov chains carries over. You will notice some weasel words creeping in, such as measurable, continuous and differentiable. These are various sorts of local regularity for functions defined on the state-space $\mathbb{R}^d$. They did not appear for Markov chains because a discrete state-space has no local structure. You might correctly guess that the proofs would require additional real analysis, relative to the corresponding results for chains, and a proper measure-theoretic basis for the probability. But, this aside, the main ideas are very similar. For further details see, for example, *Probability Theory – an analytic view* by D. W. Stroock (Cambridge University Press, 1993), or *Diffusions, Markov Processes and Martingales, Volume 1: Foundations* by L. C. G. Rogers and David Williams (Wiley, Chichester, 2nd edition 1994).

First, here is a result on recurrence and transience.

Theorem 4.4.4. Let $(B_t)_{t\geq 0}$ be a Brownian motion in $\mathbb{R}^d$.

(i) If $d = 1$, then
$$\mathbb{P}(\{t \geq 0 : B_t = 0\} \text{ is unbounded}) = 1.$$

(ii) If $d = 2$, then
$$\mathbb{P}(B_t = 0 \text{ for some } t > 0) = 0$$
but, for any $\varepsilon > 0$
$$\mathbb{P}(\{t \geq 0 : |B_t| < \varepsilon\} \text{ is unbounded}) = 1.$$

(iii) If $d = 3$, then
$$\mathbb{P}(|B_t| \to \infty \quad \text{as} \quad t \to \infty) = 1.$$

It is natural to compare this result with the facts proved in Section 1.6, that in $\mathbb{Z}$ and $\mathbb{Z}^2$ the simple symmetric random walk is recurrent, whereas in $\mathbb{Z}^3$ it is transient. The results correspond exactly in dimensions one and three. In dimension two we see the fact that for continuous state-space it makes a difference to demand returns to a point or to arbitrarily small neighbourhoods of a point. If we accept this latter notion of recurrence the correspondence extends to dimension two.

The invariant measure for Brownian motion is Lebesgue measure dx. This has infinite total mass so in dimensions one and two Brownian motion is only null recurrent. So that we can state some results for the positive recurrent case, we shall consider Brownian motion in $\mathbb{R}^d$ projected onto the torus $T^d = \mathbb{R}^d/\mathbb{Z}^d$. In dimension one this just means wrapping the line round a circle of circumference 1. The invariant measure remains Lebesgue measure but this now has total mass 1. So the projected process is positive recurrent and we can expect convergence to equilibrium and ergodic results corresponding to Theorems 1.8.3 and 1.10.2.

Theorem 4.4.5. Let $(B_t)_{t\geq 0}$ be a Brownian motion in $\mathbb{R}^d$ and let $f : \mathbb{R}^d \to \mathbb{R}$ be a continuous periodic function, so that
$$f(x + z) = f(x) \quad \text{for all } z \in \mathbb{Z}^d.$$
Then for all $x \in \mathbb{R}^d$, as $t \to \infty$, we have
$$\mathbb{E}_x[f(B_t)] \to \overline{f} = \int_{[0,1]^d} f(z) dz$$
and, moreover
$$\mathbb{P}_x\left(\frac{1}{t}\int_0^t f(B_s)ds \to \overline{f} \text{ as } t \to \infty\right) = 1.$$

The generator $\frac{1}{2}\Delta$ of Brownian motion in $\mathbb{R}^d$ reappears as it should in the following martingale characterization of Brownian motion.

Theorem 4.4.6. Let $(X_t)_{t\geq 0}$ be a continuous $\mathbb{R}^d$-valued random process. Write $(\mathcal{F}_t)_{t\geq 0}$ for the filtration of $(X_t)_{t\geq 0}$. Then the following are equivalent:

(i) $(X_t)_{t\geq 0}$ is a Brownian motion;

(ii) for all bounded functions f which are twice differentiable with bounded second derivative, the following process is a martingale:

$$M_t^f = f(X_t) - f(X_0) - \tfrac{1}{2}\int_0^t \Delta f(X_s)ds.$$

This result obviously corresponds to Theorem 4.1.2. In case you are unsure, a continuous time process $(M_t)_{t\geq 0}$ is a martingale if it is adapted to the given filtration $(\mathcal{F}_t)_{t\geq 0}$, if $\mathbb{E}|M_t| < \infty$ for all t, and

$$\mathbb{E}[(M_t - M_s)1_A] = 0$$

whenever $s \leq t$ and $A \in \mathcal{F}_s$.

We end with a result on the potentials associated with Brownian motion, corresponding very closely to Theorem 4.2.3 for Markov chains. These potentials are identical to those appearing in Newton's theory of gravity, as we remarked in Section 4.2.

Theorem 4.4.7. Let D be an open set in $\mathbb{R}^d$ with smooth boundary ∂D. Let $c: D \to [0,\infty)$ be measurable and let $f: \partial D \to [0,\infty)$ be continuous. Set

$$\phi(x) = \mathbb{E}_x\left[\int_0^T c(B_t)dt + f(X_T)1_{T<\infty}\right]$$

where T is the hitting time of ∂D. Then

(i) ϕ if finite belongs to $C^2(D) \cap C(\overline{D})$ and satisfies

$$\begin{cases} -\tfrac{1}{2}\Delta\phi = c & \text{in } D \\ \phi = f & \text{in } \partial D; \end{cases} \tag{4.12}$$

(ii) if $\psi \in C^2(D) \cap C(\overline{D})$ and satisfies

$$\begin{cases} -\tfrac{1}{2}\Delta\psi \geq c & \text{in } D \\ \psi \geq f & \text{in } \partial D \end{cases}$$

and $\psi \geq 0$, then $\psi \geq \phi$;

(iii) if $\phi(x) = \infty$ for some x, then (4.12) has no finite solution;

(iv) if $\mathbb{P}_x(T < \infty) = 1$ for all x, then (4.12) has at most one bounded solution in $C^2(D) \cap C(\overline{D})$.

5

Applications

Applications of Markov chains arise in many different areas. Some have already appeared to illustrate the theory, from games of chance to the evolution of populations, from calculating the fair price for a random reward to calculating the probability that an absent-minded professor is caught without an umbrella. In a real-world problem involving random processes you should always look for Markov chains. They are often easy to spot. Once a Markov chain is identified, there is a qualitative theory which limits the sorts of behaviour that can occur – we know, for example, that every state is either recurrent or transient. There are also good computational methods – for hitting probabilities and expected rewards, and for long-run behaviour via invariant distributions.

In this chapter we shall look at five areas of application in detail: biological models, queueing models, resource management models, Markov decision processes and Markov chain Monte Carlo. In each case our aim is to provide an introduction rather than a systematic account or survey of the field. References to books for further reading are given in each section.

5.1 Markov chains in biology

Randomness is often an appropriate model for systems of high complexity, such as are often found in biology. We have already illustrated some aspects of the theory by simple models with a biological interpretation. See Example 1.1.5 (virus), Exercise 1.1.6 (octopus), Example 1.3.4 (birth-and-death chain) and Exercise 2.5.1 (bacteria). We are now going to give

5.1 Markov chains in biology

some more examples where Markov chains have been used to model biological processes, in the study of population growth, epidemics and genetic inheritance. It should be recognised from the start that these models are simplified and somewhat stylized in order to make them mathematically tractable. Nevertheless, by providing quantitative understanding of various phenomena they can provide a useful contribution to science.

Example 5.1.1 (Branching processes)

The original branching process was considered by Galton and Watson in the 1870s while seeking a quantitative explanation for the phenomenon of the disappearance of family names, even in a growing population. Under the assumption that each male in a given family had a probability p_k of having k sons, they wished to determine the probability that after n generations an individual had no male descendents. The solution to this problem is explained below.

The basic branching process model has many applications to problems of population growth, and also to the study of chain reactions in chemistry and nuclear fission. Suppose at time $n = 0$ there is one individual, who dies and is replaced at time $n = 1$ by a random number of offspring N. Suppose, next, that these offspring also die and are themselves replaced at time $n = 2$, each independently, by a random number of further offspring, having the same distribution as N, and so on. We can construct the process by taking for each $n \in \mathbb{N}$ a sequence of independent random variables $(N_k^n)_{k \in \mathbb{N}}$, each with the same distribution as N, by setting $X_0 = 1$ and defining inductively, for $n \geq 1$

$$X_n = N_1^n + \ldots + N_{X_{n-1}}^n.$$

Then X_n gives the size of the population in the nth generation. The process $(X_n)_{n \geq 0}$ is a Markov chain on $I = \{0, 1, 2, \ldots\}$ with absorbing state 0. The case where $\mathbb{P}(N = 1) = 1$ is trivial so we exclude it. We have

$$\mathbb{P}(X_n = 0 \mid X_{n-1} = i) = \mathbb{P}(N = 0)^i$$

so if $\mathbb{P}(N = 0) > 0$ then i leads to 0, and every state $i \geq 1$ is transient. If $\mathbb{P}(N = 0) = 0$ then $\mathbb{P}(N \geq 2) > 0$, so for $i \geq 1$, i leads to j for some $j > i$, and j does not lead to i, hence i is transient in any case. We deduce that with probability 1 either $X_n = 0$ for some n or $X_n \to \infty$ as $n \to \infty$.

Further information on $(X_n)_{n \geq 0}$ is obtained by exploiting the branching structure. Consider the probability generating function

$$\phi(t) = \mathbb{E}(t^N) = \sum_{k=0}^{\infty} t^k \mathbb{P}(N = k),$$

defined for $0 \leq t \leq 1$. Conditional on $X_{n-1} = k$ we have

$$X_n = N_1^n + \ldots + N_k^n$$

so

$$\mathbb{E}(t^{X_n} \mid X_{n-1} = k) = \mathbb{E}(t^{N_1^n + \ldots + N_k^n}) = \phi(t)^k$$

and so

$$\mathbb{E}(t^{X_n}) = \sum_{k=0}^{\infty} \mathbb{E}(t^{X_n} \mid X_{n-1} = k) \mathbb{P}(X_{n-1} = k) = \mathbb{E}\big(\phi(t)^{X_{n-1}}\big).$$

Hence, by induction, we find that $\mathbb{E}(t^{X_n}) = \phi^{(n)}(t)$, where $\phi^{(n)}$ is the n-fold composition $\phi \circ \ldots \circ \phi$. In principle, this gives the entire distribution of X_n, though $\phi^{(n)}$ may be a rather complicated function. Some quantities are easily deduced: we have

$$\mathbb{E}(X_n) = \lim_{t \uparrow 1} \frac{d}{dt} \mathbb{E}(t^{X_n}) = \lim_{t \uparrow 1} \frac{d}{dt} \phi^{(n)}(t) = \Big(\lim_{t \uparrow 1} \phi'(t)\Big)^n = \mu^n,$$

where $\mu = \mathbb{E}(N)$; also

$$\mathbb{P}(X_n = 0) = \phi^{(n)}(0)$$

so, since 0 is absorbing, we have

$$q = \mathbb{P}(X_n = 0 \text{ for some } n) = \lim_{n \to \infty} \phi^{(n)}(0).$$

Now $\phi(t)$ is a convex function with $\phi(1) = 1$. Let us set $r = \inf\{t \in [0,1] : \phi(t) = t\}$, then $\phi(r) = r$ by continuity. Since ϕ is increasing and $0 \leq r$, we have $\phi(0) \leq r$ and, by induction, $\phi^{(n)}(0) \leq r$ for all n, hence $q \leq r$. On the other hand

$$q = \lim_{n \to \infty} \phi^{(n+1)}(0) = \lim_{n \to \infty} \phi(\phi^{(n)}(0)) = \phi(q)$$

so also $q \geq r$. Hence $q = r$. If $\phi'(1) > 1$ then we must have $q < 1$, and if $\phi'(1) \leq 1$ then since either $\phi'' = 0$ or $\phi'' > 0$ everywhere in $[0,1)$ we must have $q = 1$. We have shown that the population survives with positive probability if and only if $\mu > 1$, where μ is the mean of the offspring distribution.

There is a nice connection between branching processes and random walks. Suppose that in each generation we replace individuals by their offspring one at a time, so if $X_n = k$ then it takes k steps to obtain X_{n+1}.

The population size then performs a random walk $(Y_m)_{m\geq 0}$ with step distribution $N - 1$. Define stopping times $T_0 = 0$ and, for $n \geq 0$
$$T_{n+1} = T_n + Y_{T_n}.$$
Observe that $X_n = Y_{T_n}$ for all n, and since $(Y_m)_{m\geq 0}$ jumps down by at most 1 each time, $(X_n)_{n\geq 0}$ hits 0 if and only if $(Y_m)_{m\geq 0}$ hits 0. Moreover we can use the strong Markov property and a variation of the argument of Example 1.4.3 to see that, if
$$q_i = \mathbb{P}(Y_m = 0 \text{ for some } m \mid Y_0 = i)$$
then $q_i = q_1^i$ for all i and so
$$q_1 = \mathbb{P}(N = 0) + \sum_{k=1}^{\infty} q_1^i \mathbb{P}(N = i) = \phi(q_1).$$

Now each non-negative solution of this equation provides a non-negative solution of the hitting probability equations, so we deduce that q_1 is the smallest non-negative root of the equation $q = \phi(q)$, in agreement with the generating function approach.

The classic work in this area is *The Theory of Branching Processes* by T. E. Harris (Dover, New York, 1989).

Example 5.1.2 (Epidemics)

Many infectious diseases persist at a low intensity in a population for long periods. Occasionally a large number of cases arise together and form an epidemic. This behaviour is to some extent explained by the observation that the presence of a large number of infected individuals increases the risk to the rest of the population. The decline of an epidemic can also be explained by the eventual decline in the number of individuals susceptible to infection, as infectives either die or recover and are then resistant to further infection. However, these naive explanations leave unanswered many quantitative questions that are important in predicting the behaviour of epidemics.

In an idealized population we might suppose that all pairs of individuals make contact randomly and independently at a common rate, whether infected or not. For an idealized disease we might suppose that on contact with an infective, individuals themselves become infective and remain so for an exponential random time, after which they either die or recover. These two possibilities have identical consequences for the progress of the epidemic. This idealized model is obviously unrealistic, but it is the simplest mathematical model to incorporate the basic features of an epidemic.

We denote the number of susceptibles by S_t and the number of infectives by I_t. In the idealized model, $X_t = (S_t, I_t)$ performs a Markov chain on $(\mathbb{Z}^+)^2$ with transition rates

$$q_{(s,i)(s-i,i+1)} = \lambda si, \quad q_{(s,i)(s,i-1)} = \mu i$$

for some $\lambda, \mu \in (0, \infty)$. Since $S_t + I_t$ does not increase, we effectively have a finite state-space. The states $(s, 0)$ for $s \in \mathbb{Z}^+$ are all absorbing and all the other states are transient; indeed all the communicating classes are singletons. The epidemic must therefore eventually die out, and the absorption probabilities give the distribution of the number of susceptibles who escape infection. We can calculate these probabilities explicitly when $S_0 + I_0$ is small.

Of greater concern is the behaviour of an epidemic in a large population, of size N, say. Let us consider the proportions $s_t^N = S_t/N$ and $i_t^N = I_t/N$ and suppose that $\lambda = \nu/N$, where ν is independent of N. Consider now a sequence of models as $N \to \infty$ and choose $s_0^N \to s_0$ and $i_0^N \to i_0$. It can be shown that as $N \to \infty$ the process (s_t^N, i_t^N) converges to the solution (s_t, i_t) of the differential equations

$$(d/dt)s_t = -\nu s_t i_t$$
$$(d/dt)i_t = \nu s_t i_t - \mu i_t$$

starting from (s_0, i_0). Here convergence means that $\mathbb{E}[|(s_t^N, i_t^N) - (s_t, i_t)|] \to 0$ for all $t \geq 0$. We will not prove this result, but will give an example of another easier asymptotic calculation.

Consider the case where $S_0 = N-1$, $I_0 = 1$, $\lambda = 1/N$ and $\mu = 0$. This has the following interpretation: a rumour is begun by a single individual who tells it to everyone she meets; they in turn pass the rumour on to everyone they meet. We assume that each individual meets another randomly at the jump times of a Poisson process of rate 1. How long does it take until everyone knows the rumour? If i people know the rumour, then $N - i$ do not, and the rate at which the rumour is passed on is

$$q_i = i(N-i)/N.$$

The expected time until everyone knows the rumour is then

$$\sum_{i=1}^{N-1} q_i^{-1} = \sum_{i=1}^{N-1} \frac{N}{i(N-i)} = \sum_{i=1}^{N-1} \left(\frac{1}{i} + \frac{1}{N-i}\right) = 2\sum_{i=1}^{N-1} \frac{1}{i} \sim 2\log N$$

as $N \to \infty$. This is not a limit as above but, rather, an asymptotic equivalence. The fact that the expected time grows with N is related to the fact that we do not scale I_0 with N: when the rumour is known by very few or by almost all, the proportion of 'infectives' changes very slowly.

The final two examples come from population genetics. They represent an attempt to understand quantitatively the consequences of randomness in genetic inheritance. The randomness here might derive from the choice of reproducing individual, in sexual reproduction the choice of partner, or the choice of parents' alleles retained by their offspring. (The word *gene* refers to a particular chromosomal locus; the varieties of genetic material that can be present at such a locus are known as *alleles*.) This sort of study was motivated in the first place by a desire to find mathematical models of natural selection, and thereby to discriminate between various competing accounts of the process of evolution. More recently, as scientists have gained access to the genetic material itself, many more questions of a statistical nature have arisen. We emphasise that we present only the very simplest examples in a rich theory, for which we refer the interested reader to *Mathematical Population Genetics* by W. J. Ewens (Springer, Berlin, 1979).

Example 5.1.3 (Wright–Fisher model)

This is the discrete-time Markov chain on $\{0, 1, \ldots, m\}$ with transition probabilities
$$p_{ij} = \binom{m}{j} \left(\frac{i}{m}\right)^j \left(\frac{m-i}{m}\right)^{m-j}.$$
In each generation there are m alleles, some of type A and some of type a. The types of alleles in generation $n+1$ are found by choosing randomly (with replacement) from the types in generation n. If X_n denotes the number of alleles of type A in generation n, then $(X_n)_{n\geq 0}$ is a Markov chain with the above transition probabilities.

This can be viewed as a model of inheritance for a particular gene with two alleles A and a. We suppose that each individual has two genes, so the possibilities are AA, Aa and aa. Let us take m to be even with $m = 2k$. Suppose that individuals in the next generation are obtained by mating randomly chosen individuals from the current generation and that offspring inherit one allele from each parent. We have to allow that both parents may be the same, and in particular make no requirement that parents be of opposite sexes. Then if the generation n is, for example
$$AA \quad aA \quad AA \quad AA \quad aa,$$
then each gene in generation $n + 1$ is A with probability $7/10$ and a with probability $3/10$, all independent. We might, for example, get
$$aa \quad aA \quad Aa \quad AA \quad AA.$$

The structure of pairs of genes is irrelevant to the Markov chain $(X_n)_{n\geq 0}$, which simply counts the number of alleles of type A.

The communicating classes of $(X_n)_{n\geq 0}$ are $\{0\}, \{1,\ldots,m-1\}, \{m\}$. States 0 and m are absorbing and $\{1,\ldots,m-1\}$ is transient. The hitting probabilities for state m (pure AA) are given by

$$h_i = \mathbb{P}_i(X_n = m \text{ for some } n) = i/m.$$

This is obvious when one notes that $(X_n)_{n\geq 0}$ is a martingale; alternatively one can check that

$$h_i = \sum_{j=0}^{m} p_{ij} h_j.$$

According to this model, genetic diversity eventually disappears. It is known, however, that, for $p \in (0,1)$, as $m \to \infty$

$$\mathbb{E}_{pm}(T) \sim -2m\{(1-p)\log(1-p) + p\log p\}$$

where T is the hitting time of $\{0, m\}$, so in a large population diversity does not disappear quickly.

Some modifications are possible which model other aspects of genetic theory. Firstly, it may be that the three genetic types AA, Aa, aa have a relative selective advantage given by $\alpha, \beta, \gamma > 0$ respectively. This means that the probability of choosing allele A when $X_n = i$ is given by

$$\psi_i = \frac{\alpha(i/m)^2 + (1/2)\beta i(m-i)/m^2}{\alpha(i/m)^2 + \beta i(m-i)/m^2 + \gamma\big((m-i)/m\big)^2}$$

and the transition probabilities are

$$p_{ij} = \binom{m}{j} \psi_i^j (1-\psi_i)^{m-j}.$$

Secondly, we may allow genes to mutate. Suppose A mutates to a with probability u and a mutates to A with probability v. Then the probability of choosing A when $X_n = i$ is given by

$$\phi_i = \{i(1-u) + (m-i)v\}/m$$

and

$$p_{ij} = \binom{m}{j} \phi_i^j (1-\phi_i)^{m-j}.$$

With $u, v > 0$, the states 0 and m are no longer absorbing, in fact the chain is irreducible, so attention shifts from hitting probabilities to the invariant distribution π. There is an exact calculation for the mean of π: we have

$$\mu = \sum_{i=0}^{m} i\pi_i = \mathbb{E}_\pi(X_1) = \sum_{i=0}^{m} \pi_i \mathbb{E}_i(X_1)$$
$$= \sum_{i=0}^{m} m\pi_i \phi_i = \sum_{i=0}^{m} \{i(1-u) + (m-i)v\}\pi_i = (1-u)\mu + mv - v\mu$$

so that
$$\mu = mv/(u+v).$$

Example 5.1.4 (Moran model)

The Moran model is the birth-and-death chain on $\{0, 1, \ldots, m\}$ with transition probabilities

$$p_{i,i-1} = i(m-i)/m^2, \quad p_{ii} = \big(i^2 + (m-i)^2\big)/m^2, \quad p_{i,i+1} = i(m-i)/m^2.$$

Here is the genetic interpretation: a population consists of individuals of two types, a and A; we choose randomly one individual from the population at time n, and add a new individual of the same type; then we choose, again randomly, one individual from the population at time n and remove it; so we obtain the population at time $n+1$. The same individual may be chosen each time, both to give birth and to die, in which case there is no change in the make-up of the population. Now, if X_n denotes the number of type A individuals in the population at time n, then $(X_n)_{n\geq 0}$ is a Markov chain with transition matrix P.

There are some obvious differences from the Wright–Fisher model: firstly, the Moran model cannot be interpreted in terms of a species where genes come in pairs, or where individuals have more than one parent; secondly in the Moran model we only change one individual at a time, not the whole population. However, the basic Markov chain structure is the same, with communicating classes $\{0\}, \{1, \ldots, m-1\}, \{m\}$, absorbing states 0 and m and transient class $\{1, \ldots, m-1\}$. The Moran model is reversible, and, like the Wright–Fisher model, is a martingale. The hitting probabilities are given by
$$\mathbb{P}_i(X_n = m \text{ for some } n) = i/m.$$

We can also calculate explicitly the mean time to absorption
$$k_i = \mathbb{E}_i(T)$$

where T is the hitting time of $\{0, m\}$. The simplest method is first to fix j and write down equations for the mean time k_i^j spent in j, starting from i, before absorption:

$$k_i^j = \delta_{ij} + (p_{i,i-1}k_{i-1}^j + p_{ii}k_i^j + p_{i,i+1}k_{i+1}^j) \quad \text{for } i = 1, \ldots, m-1$$
$$k_0^j = k_m^j = 0.$$

Then, for $i = 1, \ldots, m-1$

$$k_{i+1}^j - 2k_i^j + k_{i-1}^j = -\delta_{ij}m^2/j(m-j)$$

so that

$$k_i^j = \begin{cases} (i/j)k_j^j & \text{for } i \leq j \\ ((m-i)/(m-j))k_j^j & \text{for } i \geq j \end{cases}$$

where k_j^j is determined by

$$\left(\frac{m-j-1}{m-j} - 2 + \frac{j-1}{j}\right) k_j^j = -\frac{m^2}{j(m-j)}$$

which gives $k_j^j = m$. Hence

$$k_i = \sum_{j=1}^{m-1} k_i^j = m \left\{ \sum_{j=1}^{i} \left(\frac{m-i}{m-j}\right) + \sum_{j=i+1}^{m-1} \frac{i}{j} \right\}.$$

As in the Wright–Fisher model, one is really interested in the case where m is large, and $i = pm$ for some $p \in (0, 1)$. Then

$$m^{-2}k_{pm} = (1-p) \sum_{j=1}^{mp} \frac{1}{m-j} + p \sum_{j=mp+1}^{m-1} \frac{1}{j} \to -(1-p)\log(1-p) - p\log p$$

as $m \to \infty$. So, as $m \to \infty$

$$\mathbb{E}_{pm}(T) \sim -m^2\{(1-p)\log(1-p) + p\log p\}.$$

For the Wright–Fisher model we claimed that

$$\mathbb{E}_{pm}(T) \sim -2m\{(1-p)\log(1-p) + p\log p\}$$

which has the same functional form in p and differs by a factor of $m/2$. This factor is partially explained by the fact that the Moran model deals with one individual at a time, whereas the Wright–Fisher model changes all m at once.

Exercises

5.1.1 Consider a branching process with immigration. This is defined, in the notation of Example 5.1.1, by
$$X_n = N_1^n + \ldots + N_{X_{n-1}}^n + I_n$$
where $(I_n)_{n \geq 0}$ is a sequence of independent $\mathbb{Z}^+$-valued random variables with common generating function $\psi(t) = \mathbb{E}(t^{I_n})$. Show that, if $X_0 = 1$, then
$$\mathbb{E}(t^{X_n}) = \phi^{(n)}(t) \prod_{k=0}^{n-1} \psi\bigl(\phi^{(k)}(t)\bigr).$$

In the case where the number of immigrants in each generation is Poisson of parameter λ, and where $\mathbb{P}(N = 0) = 1 - p$ and $\mathbb{P}(N = 1) = p$, find the long-run proportion of time during which the population is zero.

5.1.2 A species of plant comes in three genotypes AA, Aa and aa. A single plant of genotype Aa is crossed with itself, so that the offspring has genotype AA, Aa or aa with probabilities $1/4$, $1/2$ and $1/4$. How long on average does it take to achieve a pure strain, that is, AA or aa? Suppose it is desired to breed an AA plant. What should you do? How many crosses would your procedure require, on average?

5.1.3 In the Moran model we may introduce a selective bias by making it twice as likely that a type a individual is chosen to die, as compared to a type A individual. Thus in a population of size m containing i type A individuals, the probability that some type A is chosen to die is now $i/(i + 2(m - i))$. Suppose we begin with just one type A. What is the probability that eventually the whole population is of type A?

5.2 Queues and queueing networks

Queues form in many circumstances and it is important to be able to predict their behaviour. The basic mathematical model for queues runs as follows: there is a succession of customers wanting service; on arrival each customer must wait until a server is free, giving priority to earlier arrivals; it is assumed that the times between arrivals are independent random variables of the same distribution, and the times taken to serve customers are also independent random variables, of some other distribution. The main quantity of interest is the random process $(X_t)_{t\geq 0}$ recording the number of customers in the queue at time t. This is always taken to include both those being served and those waiting to be served.

In cases where inter-arrival times and service times have exponential distributions, $(X_t)_{t\geq 0}$ turns out to be a *continuous-time Markov chain*, so we can answer many questions about the queue. This is the context of our first six examples. Some further variations on queues of this type have already appeared in Exercises 3.4.1, 3.6.3, 3.7.1 and 3.7.2.

If the inter-arrival times only are exponential, an analysis is still possible, by exploiting the memorylessness of the *Poisson process* of arrivals, and a certain *discrete-time Markov chain* embedded in the queue. This is explained in the final two examples.

In each example we shall aim to describe some salient features of the queue in terms of the given data of arrival-time and service-time distributions. We shall find conditions for the stability of the queue, and in the stable case find means to compute the equilibrium distribution of queue length. We shall also look at the random times that customers spend waiting and the length of time that servers are continuously busy.

Example 5.2.1 (M/M/1 queue)

This is the simplest queue of all. The code means: *memoryless inter-arrival times/memoryless service times/one server*. Let us suppose that the inter-arrival times are exponential of parameter λ, and the service times are exponential of parameter μ. Then the number of customers in the queue $(X_t)_{t\geq 0}$ evolves as a Markov chain with the following diagram:

To see this, suppose at time 0 there are i customers in the queue, where $i > 0$. Denote by T the time taken to serve the first customer and by A the time of the next arrival. Then the first jump time J_1 is $A \wedge T$, which is exponential of parameter $\lambda + \mu$, and $X_{J_1} = i - 1$ if $T < A$, $X_{J_1} = i + 1$ if $T > A$, which events are independent of J_1, with probabilities $\mu/(\lambda+\mu)$ and $\lambda/(\lambda+\mu)$ respectively. If we condition on $J_1 = T$, then $A - J_1$ is exponential of parameter λ and independent of J_1: the time already spent waiting for an arrival is forgotten. Similarly, conditional on $J_1 = A$, $T - J_1$ is exponential of parameter μ and independent of J_1. The case where $i = 0$ is simpler as there is no serving going on. Hence, conditional on $X_{J_1} = j$, $(X_t)_{t\geq 0}$

5.2 Queues and queueing networks

begins afresh from j at time J_1. It follows that $(X_t)_{t\geq 0}$ is the claimed Markov chain. This sort of argument should by now be very familiar and we shall not spell out the details like this in later examples.

The M/M/1 queue thus evolves like a random walk, except that it does not take jumps below 0. We deduce that if $\lambda > \mu$ then $(X_t)_{t\geq 0}$ is transient, that is $X_t \to \infty$ as $t \to \infty$. Thus if $\lambda > \mu$ the *queue grows without limit* in the long term. When $\lambda < \mu$, $(X_t)_{t\geq 0}$ is positive recurrent with invariant distribution

$$\pi_i = (1 - \lambda/\mu)(\lambda/\mu)^i.$$

So when $\lambda < \mu$ the *average number of customers in the queue in equilibrium* is given by

$$\mathbb{E}_\pi(X_t) = \sum_{i=1}^{\infty} \mathbb{P}_\pi(X_t \geq i) = \sum_{i=1}^{\infty} (\lambda/\mu)^i = \lambda/(\mu - \lambda).$$

Also, the mean time to return to 0 is given by

$$m_0 = 1/(q_0 \pi_0) = \mu/\lambda(\mu - \lambda)$$

so the *mean length of time that the server is continuously busy* is given by

$$m_0 - (1/q_0) = 1/(\mu - \lambda).$$

Another quantity of interest is the *mean waiting time for a typical customer*, when $\lambda < \mu$ and the queue is in equilibrium. Conditional on finding a queue of length i on arrival, this is $(i+1)/\mu$, so the overall mean waiting time is

$$\mathbb{E}_\pi(X_t + 1)/\mu = 1/(\mu - \lambda).$$

A rough check is available here as we can calculate in two ways the expected total time spent in the queue over an interval of length t: either we multiply the average queue length by t, or we multiply the mean waiting time by the expected number of customers λt. Either way we get $\lambda t/\mu - \lambda$. The first calculation is exact but we have not fully justified the second.

Thus, once the queue size is identified as a Markov chain, its behaviour is largely understood. Even in more complicated examples where exact calculation is limited, once the Markovian character of the queue is noted we know what sort of features to look for – transience and recurrence, convergence to equilibrium, long-run averages, and so on.

Example 5.2.2 (M/M/s queue)

This is a variation on the last example where there is one queue but there are s servers. Let us assume that the arrival rate is λ and the service rate by each server is μ. Then if i servers are occupied, the first service is completed at the minimum of i independent exponential times of parameter μ. The first service time is therefore exponential of parameter $i\mu$. The total service rate increases to a maximum $s\mu$ when all servers are working. We emphasise that the queue size includes those customers who are currently being served. By an argument similar to the preceding example, the queue size $(X_t)_{t\geq 0}$ performs a Markov chain with the following diagram:

$$
\begin{array}{c}
\xrightarrow{\lambda} \; \xleftarrow{\mu} \; \xrightarrow{\lambda} \; \xleftarrow{2\mu} \; \xrightarrow{\lambda} \; \cdots \; \xleftarrow{s\mu} \; \xrightarrow{\lambda} \; \xleftarrow{s\mu} \; \xrightarrow{\lambda} \; \cdots \\
0 \quad\quad 1 \quad\quad 2 \quad\quad\quad\quad s \quad\quad s+1
\end{array}
$$

So this time we obtain a birth-and-death chain. It is transient in the case $\lambda > s\mu$ and otherwise recurrent. To find an invariant measure we look at the detailed balance equations

$$\pi_i q_{i,i+1} = \pi_{i+1} q_{i+1,i}.$$

Hence
$$\pi_i/\pi_0 = \begin{cases} (\lambda/\mu)^i/i! & \text{for } i = 0, 1, \ldots, s \\ (\lambda/\mu)^i/(s^{i-s}s!) & \text{for } i = s+1, s+2, \ldots. \end{cases}$$

The queue is therefore positive recurrent when $\lambda < s\mu$. There are two cases when the invariant distribution has a particularly nice form: when $s = 1$ we are back to Example 5.2.1 and the invariant distribution is geometric of parameter λ/μ:

$$\pi_i = (1 - \lambda/\mu)(\lambda/\mu)^i.$$

When $s = \infty$ we normalize π by taking $\pi_0 = e^{-\lambda/\mu}$ so that

$$\pi_i = e^{-\lambda/\mu}(\lambda/\mu)^i/i!$$

and the invariant distribution is Poisson of parameter λ/μ.

The number of arrivals by time t is a Poisson process of rate λ. Each arrival corresponds to an increase in X_t, and each departure to a decrease. Let us suppose that $\lambda < s\mu$, so there is an invariant distribution, and consider the queue in equilibrium. The detailed balance equations hold and $(X_t)_{t\geq 0}$ is non-explosive, so by Theorem 3.7.3 for any $T > 0$, $(X_t)_{0\leq t\leq T}$

5.2 Queues and queueing networks

and $(X_{T-t})_{0 \leq t \leq T}$ have the same law. It follows that, in equilibrium, the number of departures by time t is also a Poisson process of rate λ. This is slightly counter-intuitive, as one might imagine that the departure process runs in fits and starts depending on the number of servers working. Instead, it turns out that the process of departures, in equilibrium, is just as regular as the process of arrivals.

Example 5.2.3 (Telephone exchange)

A variation on the M/M/s queue is to turn away customers who cannot be served immediately. This might serve as a simple model for a telephone exchange, where the maximum number of calls that can be connected at once is s: when the exchange is full, additional calls are lost. The maximum queue size or *buffer size* is s and we get the following modified Markov chain diagram:

We can find the invariant distribution of this finite Markov chain by solving the detailed balance equations, as in the last example. This time we get a *truncated Poisson distribution*

$$\pi_i = \frac{(\lambda/\mu)^i}{i!} \bigg/ \sum_{j=0}^{s} \frac{(\lambda/\mu)^j}{j!}.$$

By the ergodic theorem, the long-run proportion of time that the exchange is full, and hence the long-run proportion of calls that are lost, is given by

$$\pi_s = \frac{(\lambda/\mu)^s}{s!} \bigg/ \sum_{j=0}^{s} \frac{(\lambda/\mu)^j}{j!}.$$

This is known as *Erlang's formula*. Compare this example with the bus maintenance problem in Exercise 3.7.1.

Example 5.2.4 (Queues in series)

Suppose that customers have two service requirements: they arrive as a Poisson process of rate λ to be seen first by server A, and then by server

B. For simplicity we shall assume that the service times are independent exponentials of parameters α and β respectively. What is the average queue length at B?

Let us denote the queue length at A by $(X_t)_{t\geq 0}$ and that by B by $(Y_t)_{t\geq 0}$. Then $(X_t)_{t\geq 0}$ is simply an M/M/1 queue. If $\lambda > \alpha$, then $(X_t)_{t\geq 0}$ is transient so there is eventually always a queue at A and departures form a Poisson process of rate α. If $\lambda < \alpha$, then, by the reversibility argument of Example 5.2.2, the process of departures from A is Poisson of rate λ, *provided queue A is in equilibrium*. The question about queue length at B is not precisely formulated: it does not specify that the queues should be in equilibrium; indeed if $\lambda \geq \alpha$ there is no equilibrium. Nevertheless, we hope you will agree to treat arrivals at B as a Poisson process of rate $\alpha \wedge \lambda$. Then, by Example 5.2.1, the average queue length at B when $\alpha \wedge \lambda < \beta$, in equilibrium, is given by $(\alpha \wedge \lambda)/(\beta - (\alpha \wedge \lambda))$. If, on the other hand, $\alpha \wedge \lambda > \beta$, then $(Y_t)_{t\geq 0}$ is transient so the queue at B grows without limit.

There is an equilibrium for both queues if $\lambda < \alpha$ and $\lambda < \beta$. The fact that in equilibrium the output from A is Poisson greatly simplifies the analysis of the two queues in series. For example, the average time taken by one customer to obtain both services is given by

$$1/(\alpha - \lambda) + 1/(\beta - \lambda).$$

Example 5.2.5 (Closed migration process)

Consider, first, a single particle in a finite state-space I which performs a Markov chain with irreducible Q-matrix Q. We know there is a unique invariant distribution π. We may think of the holding times of this chain as service times, by a single server at each node $i \in I$.

Let us suppose now that there are N particles in the state-space, which move as before except that they must queue for service at every node. If we do not care to distinguish between the particles, we can regard this as a new process $(X_t)_{t\geq 0}$ with state-space $\widetilde{I} = \mathbb{N}^I$, where $X_t = (n_i : i \in I)$ if at time t there are n_i particles at state i. In fact, this new process is also a Markov chain. To describe its Q-matrix $\widetilde{Q}$ we define a function $\delta_i : \widetilde{I} \to \widetilde{I}$ by

$$(\delta_i n)_j = n_j + \delta_{ij}.$$

Thus δ_i adds a particle at i. Then for $i \neq j$ the non-zero transition rates are given by

$$\widetilde{q}(\delta_i n, \delta_j n) = q_{ij}, \qquad n \in \widetilde{I}, \quad i, j \in I.$$

5.2 Queues and queueing networks

Observe that we can write the invariant measure equation $\pi Q = 0$ in the form
$$\pi_i \sum_{j \neq i} q_{ij} = \sum_{j \neq i} \pi_j q_{ji}. \tag{5.1}$$

For $n = (n_i : i \in I)$ we set
$$\widetilde{\pi}(n) = \prod_{i \in I} \pi_i^{n_i}.$$

Then
$$\widetilde{\pi}(\delta_i n) \sum_{j \neq i} \widetilde{q}(\delta_i n, \delta_j n) = \prod_{k \in I} \pi_k^{n_k} \left(\pi_i \sum_{j \neq i} q_{ji} \right)$$
$$= \prod_{k \in I} \pi_k^{n_k} \left(\sum_{j \neq i} \pi_j q_{ji} \right)$$
$$= \sum_{j \neq i} \widetilde{\pi}(\delta_j n) \widetilde{q}(\delta_j n, \delta_i n).$$

Given $m \in \widetilde{I}$ we can put $m = \delta_i n$ in the last identity whenever $m_i \geq 1$. On summing the resulting equations we obtain
$$\widetilde{\pi}(m) \sum_{n \neq m} \widetilde{q}(m, n) = \sum_{n \neq m} \widetilde{\pi}(n) \widetilde{q}(n, m)$$

so $\widetilde{\pi}$ is an invariant measure for $\widetilde{Q}$. The total number of particles is conserved so $\widetilde{Q}$ has communicating classes
$$C_N = \left\{ n \in \widetilde{I} : \sum_{i \in I} n_i = N \right\}$$

and the unique invariant distribution for the N-particle system is given by normalizing $\widetilde{\pi}$ restricted to C_N.

Example 5.2.6 (Open migration process)

We consider a modification of the last example where new customers, or particles, arrive at each node $i \in I$ at rate λ_i. We suppose also that customers receiving service at node i leave the network at rate μ_i. Thus customers enter the network, move from queue to queue according to a Markov chain and eventually leave, rather like a shopping centre. This model includes the closed system of the last example and also the queues in series of Example 5.2.4. Let $X_t = (X_t^i : i \in I)$, where X_t^i denotes the number of customers at node i at time t. Then $(X_t)_{t \geq 0}$ is a Markov chain in $\widetilde{I} = \mathbb{N}^I$ and the non-zero transition rates are given by

$$\widetilde{q}(n, \delta_i n) = \lambda_i, \quad \widetilde{q}(\delta_i n, \delta_j n) = q_{ij}, \quad \widetilde{q}(\delta_j n, n) = \mu_j$$

for $n \in \widetilde{I}$ and distinct states $i, j \in I$. We shall assume that $\lambda_i > 0$ for some i and $\mu_j > 0$ for some j; then $\widetilde{Q}$ is irreducible on $\widetilde{I}$.

The system of equations (5.1) for an invariant measure is replaced here by

$$\pi_i \left(\mu_i + \sum_{j \neq i} q_{ij} \right) = \lambda_i + \sum_{j \neq i} \pi_j q_{ji}.$$

This system has a unique solution, with $\pi_i > 0$ for all i. This may be seen by considering the invariant distribution for the extended Q-matrix $\overline{Q}$ on $I \cup \{\partial\}$ with off-diagonal entries

$$\overline{q}_{\partial j} = \lambda_j, \quad \overline{q}_{ij} = q_{ij}, \quad \overline{q}_{i\partial} = \mu_i.$$

On summing the system over $i \in I$ we find

$$\sum_{i \in I} \pi_i \mu_i = \sum_{i \in I} \lambda_i.$$

As in the last example, for $n = (n_i : i \in I)$ we set

$$\widetilde{\pi}(n) = \prod_{i \in I} \pi_i^{n_i}.$$

Transitions from $m \in \widetilde{I}$ may be divided into those where a new particle is added and, for each $i \in I$ with $m_i \geq 1$, those where a particle is moved from i to somewhere else. We have, for the first sort of transition

$$\widetilde{\pi}(m) \sum_{j \in I} \widetilde{q}(m, \delta_j m) = \widetilde{\pi}(m) \sum_{j \in I} \lambda_j$$

$$= \widetilde{\pi}(m) \sum_{j \in I} \pi_j \mu_j = \sum_{j \in I} \widetilde{\pi}(\delta_j m) \widetilde{q}(\delta_j m, m)$$

and for the second sort

$$\widetilde{\pi}(\delta_i n) \left(\widetilde{q}(\delta_i n, n) + \sum_{j \neq i} \widetilde{q}(\delta_i n, \delta_j n) \right)$$

$$= \prod_{k \in I} \pi_k^{n_k} \left(\pi_i (\mu_i + \sum_{j \neq i} q_{ij}) \right)$$

$$= \prod_{k \in I} \pi_k^{n_k} \left(\lambda_i + \sum_{j \neq i} \pi_j q_{ji} \right)$$

$$= \widetilde{\pi}(n) \widetilde{q}(n, \delta_i n) + \sum_{j \neq i} \widetilde{\pi}(\delta_j n) \widetilde{q}(\delta_j n, \delta_i n).$$

On summing these equations we obtain

$$\widetilde{\pi}(m) \sum_{n \neq m} \widetilde{q}(m,n) = \sum_{n \neq m} \widetilde{\pi}(n) \widetilde{q}(n,m)$$

so $\widetilde{\pi}$ is an invariant measure for Q. If $\pi_i < 1$ for all i then $\widetilde{\pi}$ has finite total mass $\prod_{i \in I}(1-\pi_i)$, otherwise the total mass if infinite. Hence, $\widetilde{Q}$ is positive recurrent if and only if $\pi_i < 1$ for all i, and in that case, in equilibrium, the individual queue lengths $(X_t^i : i \in I)$ are *independent* geometric random variables with

$$\mathbb{P}(X_t^i = n_i) = (1-\pi_i)\pi_i^{n_i}.$$

Example 5.2.7 (M/G/1 queue)

As we argued in Section 2.4, the Poisson process is the natural probabilistic model for any uncoordinated stream of discrete events. So we are often justified in assuming that arrivals to a queue form a Poisson process. In the preceding examples we also assumed an exponential service-time distribution. This is desirable because it makes the queue size into a continuous-time Markov chain, but it is obviously inappropriate in many real-world examples. The service requirements of customers and the duration of telephone calls have observable distributions which are generally not exponential. A better model in this case is the M/G/1 queue, where G indicates that the service-time distribution is general.

We can characterize the distribution of a service time T by its distribution function

$$F(t) = \mathbb{P}(T \leq t),$$

or by its Laplace transform

$$L(w) = \mathbb{E}(e^{-wT}) = \int_0^\infty e^{-wt} dF(t).$$

(The integral written here is the Lebesgue–Stieltjes integral: when T has a density function $f(t)$ we can replace $dF(t)$ by $f(t)dt$.) Then the mean service time μ is given by

$$\mu = \mathbb{E}(T) = -L'(0+).$$

To analyse the M/G/1 queue, we consider the queue size X_n immediately following the nth departure. Then

$$X_{n+1} = X_n + Y_{n+1} - 1_{X_n > 0} \qquad (5.2)$$

where Y_n denotes the number of arrivals during the nth service time. The case where $X_n = 0$ is different because then we get an extra arrival before the $(n+1)$th service time begins. By the Markov property of the Poisson process, $Y_1, Y_2, \ldots$ are independent and identically distributed, so $(X_n)_{n\geq 0}$ is a *discrete-time Markov chain*. Indeed, except for visits to 0, $(X_n)_{n\geq 0}$ behaves as a random walk with jumps $Y_n - 1$.

Let T_n denote the nth service time. Then, conditional on $T_n = t$, Y_n is Poisson of parameter λt. So

$$\mathbb{E}(Y_n) = \int_0^\infty \lambda t \, dF(t) = \lambda \mu$$

and, indeed, we can compute the probability generating function

$$A(z) = \mathbb{E}(z^{Y_n}) = \int_0^\infty \mathbb{E}(z^{Y_n} \mid T_n = t) dF(t)$$
$$= \int_0^\infty e^{-\lambda t(1-z)} dF(t) = L(\lambda(1-z)).$$

Set $\rho = \mathbb{E}(Y_n) = \lambda \mu$. We call ρ the *service intensity*. Let us suppose that $\rho < 1$. We have

$$X_n = X_0 + (Y_1 + \ldots + Y_n) - n + Z_n$$

where Z_n denotes the number of visits of X_n to 0 before time n. So

$$\mathbb{E}(X_n) = \mathbb{E}(X_0) - n(1-\rho) + \mathbb{E}(Z_n).$$

Take $X_0 = 0$, then, since $X_n \geq 0$, we have for all n

$$0 < 1 - \rho \leq \mathbb{E}(Z_n/n).$$

By the ergodic theorem we know that, as $n \to \infty$

$$\mathbb{E}(Z_n/n) \to 1/m_0$$

where m_0 is the mean return time to 0. Hence

$$m_0 \leq 1/(1-\rho) < \infty$$

showing that $(X_n)_{n\geq 0}$ is positive recurrent.

Suppose now that we start $(X_n)_{n\geq 0}$ with its equilibrium distribution π. Set

$$G(z) = \mathbb{E}(z^{X_n}) = \sum_{i=0}^\infty \pi_i z^i$$

then
$$zG(z) = \mathbb{E}(z^{X_{n+1}+1}) = \mathbb{E}(z^{X_n+Y_{n+1}+1_{X_n=0}})$$
$$= \mathbb{E}(z^{Y_{n+1}})\left(\pi_0 z + \sum_{i=1}^{\infty}\pi_i z^i\right)$$
$$= A(z)\bigl(\pi_0 z + G(z) - \pi_0\bigr)$$

so
$$\bigl(A(z) - z\bigr)G(z) = \pi_0 A(z)(1-z). \tag{5.3}$$

By l'Hôpital's rule, as $z \uparrow 1$
$$\bigl(A(z) - z\bigr)/(1-z) \to 1 - A'(1-) = 1 - \rho.$$

Since $G(1) = 1 = A(1)$, we must therefore have $\pi_0 = 1 - \rho$, $m_0 = 1/(1-\rho)$ and
$$G(z) = (1-\rho)(1-z)A(z)/\bigl(A(z) - z\bigr).$$

Since A is given explicitly in terms of the service-time distribution, we can now obtain, in principle, the full equilibrium distribution. The fact that generating functions work well here is due to the additive structure of (5.2).

To obtain the *mean queue length* we differentiate (5.3)
$$\bigl(A(z) - z\bigr)G'(z) = \bigl(A'(z) - 1\bigr)G(z) = (1-\rho)\bigl\{A'(z)(1-z) - A(z)\bigr\},$$

then substitute for $G(z)$ to obtain
$$G'(z) = (1-\rho)A'(z)\frac{(1-z)}{\bigl(A(z)-z\bigr)} - (1-\rho)A(z)\frac{\bigl\{\bigl(A'(z)-1\bigr)(1-z) + A(z)-z\bigr\}}{\bigl(A(z)-z\bigr)^2}.$$

By l'Hôpital's rule:
$$\lim_{z\uparrow 1}\frac{\bigl(A'(z)-1\bigr)(1-z) + A(z)-z}{\bigl(A(z)-z\bigr)^2} = \lim_{z\uparrow 1}\frac{A''(z)(1-z)}{2\bigl(A'(z)-1\bigr)\bigl(A(z)-z\bigr)} = \frac{-A''(1-)}{2(1-\rho)^2}.$$

Hence
$$\mathbb{E}(X_n) = G'(1-) = \rho + A''(1-)/2(1-\rho)$$
$$= \rho + \lambda^2 L''(0+)/2(1-\rho) = \rho + \lambda^2 \mathbb{E}(T^2)/2(1-\rho).$$

In the case of the M/M/1 queue $\rho = \lambda/\mu$, $\mathbb{E}(T^2) = 2/\mu^2$ and $\mathbb{E}(X_n) = \rho/(1-\rho) = \lambda/(\mu-\lambda)$, as we found in Example 5.2.1.

We shall use generating functions to study two more quantities of interest – the queueing time of a typical customer and the busy periods of the server.

Consider the queue $(X_n)_{n\in\mathbb{Z}}$ in equilibrium. Suppose that the customer who leaves at time 0 has spent time Q queueing to be served, and time T being served. Then, conditional on $Q+T=t$, X_0 is Poisson of parameter λt, since the customers in the queue at time 0 are precisely those who arrived during the queueing and service times of the departing customer. Hence
$$G(z) = \mathbb{E}\big(e^{-\lambda(Q+T)(1-z)}\big) = M\big(\lambda(1-z)\big)L\big(\lambda(1-z)\big)$$
where M is the Laplace transform
$$M(w) = \mathbb{E}(e^{-wQ}).$$

On substituting for $G(z)$ we obtain the formula
$$M(w) = (1-\rho)w\big/\big(w - \lambda\big(1 - L(w)\big)\big).$$

Differentiation and l'Hôpital's rule, as above, lead to a formula for the *mean queueing time*
$$\mathbb{E}(Q) = -M'(0+) = \frac{\lambda L''(0+)}{2\big(1+\lambda L'(0+)\big)^2} = \frac{\lambda \mathbb{E}(T^2)}{2(1-\rho)}.$$

We now turn to the busy period S. Consider the Laplace transform
$$B(w) = \mathbb{E}(e^{-wS}).$$

Let T denote the service time of the first customer in the busy period. Then conditional on $T=t$, we have
$$S = t + S_1 + \ldots + S_N,$$
where N is the number of customers arriving while the first customer is served, which is Poisson of parameter λt, and where $S_1, S_2, \ldots$ are independent, with the same distribution as S. Hence
$$B(w) = \int_0^\infty \mathbb{E}(e^{-wS} \mid T=t) dF(t)$$
$$= \int_0^\infty e^{-wt} e^{-\lambda t(1-B(w))} dF(t) = L\big(w + \lambda\big(1 - B(w)\big)\big).$$

5.2 Queues and queueing networks

Although this is an implicit relation for $B(w)$, we can obtain moments by differentiation:

$$\mathbb{E}(S) = -B'(0+) = -L'(0+)\big(1 - \lambda B'(0+)\big) = \mu\big(1 + \lambda\mathbb{E}(S)\big)$$

so the *mean length of the busy period* is given by

$$\mathbb{E}(S) = \mu/(1-\rho).$$

Example 5.2.8 (M/G/∞ queue)

Arrivals at this queue form a Poisson process, of rate λ, say. Service times are independent, with a common distribution function $F(t) = \mathbb{P}(T \leq t)$. There are infinitely many servers, so all customers in fact receive service at once. The analysis here is simpler than in the last example because customers do not interact. Suppose there are no customers at time 0. What, then, is the distribution of the number X_t being served at time t?

The number N_t of arrivals by time t is a Poisson random variable of parameter λt. We condition on $N_t = n$ and label the times of the n arrivals randomly by $A_1, \ldots, A_n$. Then, by Theorem 2.4.6, $A_1, \ldots, A_n$ are independent and uniformly distributed on the interval $[0,t]$. For each of these customers, service is incomplete at time t with probability

$$p = \frac{1}{t}\int_0^t \mathbb{P}(T > s)ds = \frac{1}{t}\int_0^t \big(1 - F(s)\big)ds.$$

Hence, conditional on $N_t = n$, X_t is binomial of parameters n and p. Then

$$\mathbb{P}(X_t = k) = \sum_{n=0}^{\infty} \mathbb{P}(X_t = k \mid N_t = n)\mathbb{P}(N_t = n)$$

$$= \sum_{n=k}^{\infty} \binom{n}{k} p^k (1-p)^{n-k} e^{-\lambda t}(\lambda t)^n/n!$$

$$= e^{-\lambda t}(\lambda pt)^k/k! \sum_{n=k}^{\infty} \big(\lambda(1-p)t\big)^{n-k}/(n-k)!$$

$$= e^{-\lambda pt}(\lambda pt)^k/k!$$

So we have shown that X_t is Poisson of parameter

$$\lambda \int_0^t \big(1 - F(s)\big)ds.$$

Recall that

$$\int_0^\infty \big(1 - F(s)\big)ds = \int_0^\infty \mathbb{E}(1_{T>t})dt = \mathbb{E}\int_0^\infty 1_{T>t}dt = \mathbb{E}(T).$$

Hence if $\mathbb{E}(T) < \infty$, the queue size has a limiting distribution, which is Poisson of parameter $\lambda\mathbb{E}(T)$.

For further reading see *Reversibility and Stochastic Networks* by F. P. Kelly (Wiley, Chichester, 1978).

5.3 Markov chains in resource management

Management decisions are always subject to risk because of the uncertainty of future events. If one can quantify that risk, perhaps on the basis of past experience, then the determination of the best action will rest on the calculation of probabilities, often involving a Markov chain. Here we present some examples involving the management of a resource: either the stock in a warehouse, or the water in a reservoir, or the reserves of an insurance company. See also Exercise 3.7.1 on the maintenance of unreliable equipment. The statistical problem of estimating transition rates for Markov chains has already been discussed in Section 1.10.

Example 5.3.1 (Restocking a warehouse)

A warehouse has a capacity of c units of stock. In each time period n, there is a demand for D_n units of stock, which is met if possible. We denote the residual stock at the end of period n by X_n. The warehouse manager restocks to capacity for the beginning of period $n+1$ whenever $X_n \leq m$, for some threshold m. Thus $(X_n)_{n \geq 0}$ satisfies

$$X_{n+1} = \begin{cases} (c - D_{n+1})^+ & \text{if } X_n \leq m \\ (X_n - D_{n+1})^+ & \text{if } m < X_n \leq c. \end{cases}$$

Let us assume that $D_1, D_2, \ldots$ are independent and identically distributed; then $(X_n)_{n \geq 0}$ is a Markov chain, and, excepting some peculiar demand structures, is irreducible on $\{0, 1, \ldots, c\}$. Hence $(X_n)_{n \geq 0}$ has a unique invariant distribution π which determines the long-run proportion of time in each state. Given that $X_n = i$, the expected unmet demand in period $n+1$ is given by

$$u_i = \begin{cases} \mathbb{E}\big((D - c)^+\big) & \text{if } i \leq m \\ \mathbb{E}\big((D - i)^+\big) & \text{if } m < i \leq c. \end{cases}$$

5.3 Markov chains in resource management

Hence the long-run proportion of demand that is unmet is

$$u(m) = \sum_{i=0}^{c} \pi_i u_i.$$

The long-run frequency of restocking is given by

$$r(m) = \sum_{i=0}^{m} \pi_i.$$

Now as m increases, $u(m)$ decreases and $r(m)$ increases. The warehouse manager may want to compute these quantities in order to optimize the long-run cost

$$ar(m) + bu(m)$$

where a is the cost of restocking and b is the profit per unit.

There is no general formula for π, but once the distribution of the demand is known, it is a relatively simple matter to write down the $(c+1) \times (c+1)$ transition matrix P for $(X_n)_{n \geq 0}$ and solve $\pi P = \pi$ subject to $\sum_{i=0}^{c} \pi_i = 1$. We shall discuss in detail a special case where the calculations work out nicely.

Suppose that the capacity $c = 3$, so possible threshold values are $m = 0, 1, 2$. Suppose that the profit per unit $b = 1$, and that the demand satisfies

$$\mathbb{P}(D \geq i) = 2^{-i} \quad \text{for } i = 0, 1, 2, \ldots.$$

Then

$$\mathbb{E}\big((D-i)^+\big) = \sum_{k=1}^{\infty} \mathbb{P}\big((D-i)^+ \geq k\big) = \sum_{k=1}^{\infty} \mathbb{P}(D \geq i+k) = 2^{-i}.$$

The transition matrices for $m = 0, 1, 2$ are given, respectively, by

$$\begin{pmatrix} 1/8 & 1/8 & 1/4 & 1/2 \\ 1/2 & 1/2 & 0 & 0 \\ 1/4 & 1/4 & 1/2 & 0 \\ 1/8 & 1/8 & 1/4 & 1/2 \end{pmatrix} \begin{pmatrix} 1/8 & 1/8 & 1/4 & 1/2 \\ 1/8 & 1/8 & 1/4 & 1/2 \\ 1/4 & 1/4 & 1/2 & 0 \\ 1/8 & 1/8 & 1/4 & 1/2 \end{pmatrix} \begin{pmatrix} 1/8 & 1/8 & 1/4 & 1/2 \\ 1/8 & 1/8 & 1/4 & 1/2 \\ 1/8 & 1/8 & 1/4 & 1/2 \\ 1/8 & 1/8 & 1/4 & 1/2 \end{pmatrix}$$

with invariant distributions

$$(1/4, 1/4, 1/4, 1/4), \quad (1/6, 1/6, 1/3, 1/3), \quad (1/8, 1/8, 1/4, 1/2).$$

Hence

$$u(0) = 1/4, \quad u(1) = 1/6, \quad u(2) = 1/8$$

and
$$r(0) = 1/4, \quad r(1) = 1/3, \quad r(2) = 1/2.$$

Therefore, to minimize the long-run cost $ar(m) + u(m)$ we should take
$$m = \begin{cases} 2 & \text{if } a \leq 1/4 \\ 1 & \text{if } 1/4 < a \leq 1 \\ 0 & \text{if } 1 < a. \end{cases}$$

Example 5.3.2 (Reservoir model – discrete time)

We are concerned here with a storage facility, for example a reservoir, of finite capacity c. In each time period n, A_n units of resource are available to enter the facility and B_n units are drawn off. When the reservoir is full, surplus water is lost. When the reservoir is empty, no water can be supplied. We assume that newly available resources cannot be used in the current time period. Then the quantity of water X_n in the reservoir at the end of period n satisfies
$$X_{n+1} = \bigl((X_n - B_{n+1})^+ + A_{n+1}\bigr) \wedge c.$$

If we assume that A_n, B_n and c are integer-valued and that $A_1, A_2, \ldots$ are independent and identically distributed, likewise $B_1, B_2, \ldots$, then $(X_n)_{n\geq 0}$ is a Markov chain on $\{0, 1, \ldots, c\}$, whose transition probabilities may be deduced from the distributions of A_n and B_n. Hence we know that the long-run behaviour of $(X_n)_{n\geq 0}$ is controlled by its unique invariant distribution π, assuming irreducibility. For example, the long-run proportion of time that the reservoir is empty is simply π_0. So we would like to calculate π.

A simplifying assumption which makes some calculations possible is to assume that consumption in each period is constant, and that our units are chosen to make this constant 1. Then the infinite capacity model satisfies a recursion similar to the M/G/1 queue:
$$X_{n+1} = (X_n - 1)^+ + A_{n+1}.$$

Hence, by the argument used in Example 5.2.7, if $\mathbb{E}(A_n) < 1$, then $(X_n)_{n\geq 0}$ is positive recurrent and the invariant distribution π satisfies
$$\sum_{i=0}^{\infty} \pi_i z^i = (1 - \mathbb{E}A_n)(1-z)A(z)/\bigl(A(z) - z\bigr)$$

where $A(z) = \mathbb{E}(z^{A_n})$. In fact, whether or not $\mathbb{E}(A_n) < 1$, the equation
$$\sum_{i=0}^{\infty} \nu_i z^i = (1-z)A(z)/\bigl(A(z) - z\bigr)$$

serves to define a solution to the infinite system of linear equations

$$\nu_0 = \nu_0(a_0 + a_1) + \nu_1 a_0$$

$$\nu_i = \nu_{i+1} a_0 + \sum_{j=0}^{i} \nu_j a_{i-j+1}, \quad \text{for } i \geq 1$$

where $a_i = \mathbb{P}(A_n = i)$.

Note that $(X_n)_{n \geq 0}$ can only enter $\{0, 1, \ldots, c\}$ through c. Hence, by the strong Markov property, $(X_n)_{n \geq 0}$ observed whilst in $\{0, 1, \ldots, c\}$ is simply the finite-capacity model. In the case where $\mathbb{E}(A_n) < 1$, we can deduce for the finite-capacity model that the long-run proportion of time in state i is given by $\nu_i/(\nu_0 + \ldots + \nu_c)$. In fact, this is true in general as the equilibrium equations for the finite-capacity model coincide with those for ν up to level $c - 1$, and the level c equation is redundant.

In reality, it is to be hoped that, in the long run, supply will exceed demand, which is true if $\mathbb{E}(A_n) > 1$. Then $(X_n)_{n \geq 0}$ is transient, so ν must have infinite total mass. The problem faced by the water company is to keep the long-run proportion of time $\pi_0(c)$ that the reservoir is empty below a certain acceptable fraction, $\varepsilon > 0$ say. Hence c should be chosen large enough to make

$$\nu_o/(\nu_0 + \ldots + \nu_c) < \varepsilon$$

which is always possible in the transient case.

Example 5.3.3 (Reservoir model – continuous time)

Consider a reservoir model where fresh water arrives at the times of a Poisson process of rate λ. The quantities of water $S_1, S_2, \ldots$ arriving each time are assumed independent and identically distributed. We assume that there is a continuous demand for water of rate 1. For a reservoir of infinite capacity, the quantity of water held $(W_t)_{t \geq 0}$ is just the stored work in an M/G/1 queue with the same arrival times and service times $S_1, S_2, \ldots$. The periods when the reservoir is empty correspond to idle periods of the queue. Hence in the positive recurrent case where $\lambda \mathbb{E}(S_n) < 1$, the long-run proportion of time that the reservoir is empty is given by $\mathbb{E}(S_n)/(1 - \lambda \mathbb{E}(S_n))$. Note that $(W_t)_{t \geq 0}$ can enter $[0, c]$ only through c. As in the preceding example we can obtain the finite capacity model by observing $(W_t)_{t \geq 0}$ whilst in $[0, c]$, but we shall not pursue this here.

The next example is included, in part, because it illustrates a surprising and powerful connection between *reflected random walks* and the *maxima of random walks*, which we now explain. Let $X_1, X_2, \ldots$ denote a sequence of independent, identically distributed random variables. Set $S_n = X_1 + \ldots + X_n$ and define $(Z_n)_{n \geq 0}$ by $Z_0 = 0$ and

$$Z_{n+1} = (Z_n + X_{n+1})^+.$$

Then, by induction, we have

$$Z_n = \max\{0, X_n, X_{n-1} + X_n, \ldots, X_1 + \ldots + X_n\}$$

so Z_n *has the same distribution as* M_n *where*

$$M_n = \max\{0, X_1, X_1 + X_2, \ldots, X_1 + \ldots + X_n\} = \max_{m \leq n} S_m.$$

Example 5.3.4 (Ruin of an insurance company)

An insurance company receives premiums continuously at a constant rate. We choose units making this rate 1. The company pays claims at the times of a Poisson process of rate λ, the claims $Y_1, Y_2, \ldots$ being independent and identically distributed. Set $\rho = \lambda \mathbb{E}(Y_1)$ and assume that $\rho < 1$. Then in the long run the company can expect to make a profit of $1 - \rho$ per unit time. However, there is a danger that large claims early on will ruin the company even though the long-term trend is good.

Denote by S_n the cumulative net loss following the nth claim. Thus $S_n = X_1 + \ldots + X_n$, where $X_n = Y_n - T_n$ and T_n is the nth inter-arrival time. By the strong law of large numbers

$$S_n/n \to \mathbb{E}(Y_n) - 1/\lambda < 0$$

as $n \to \infty$. The maximum loss that the company will have to sustain is

$$M = \lim_{n \to \infty} M_n$$

where

$$M_n = \max_{m \leq n} S_m.$$

By the argument given above, M_n has the same distribution as Z_n, where $Z_0 = 0$ and

$$Z_{n+1} = (Z_n + Y_n - T_n)^+.$$

But Z_n is the queueing time of the nth customer in the M/G/1 queue with inter-arrival times T_n and service times Y_n. We know by Example 5.2.7 that the queue-length distribution converges to equilibrium. Hence, so does the queueing-time distribution. Also by Example 5.2.7, we know the Laplace transform of the equilibrium queueing-time distribution. Hence

$$\mathbb{E}(e^{-wM}) = (1-\rho)w/\Big(w - \lambda\big(1 - \mathbb{E}(e^{-wY_1})\big)\Big).$$

The probability of eventual bankruptcy is $\mathbb{P}(M > a)$, where a denotes the initial value of the company's assets. In principle, this may now be obtained by inverting the Laplace transform.

5.4 Markov decision processes

In many contexts costs are incurred at a rate determined by some process which may best be modelled as a Markov chain. We have seen in Section 1.10 and Section 4.2 how to calculate in these circumstances the long-run average cost or the expected total cost. Suppose now that we are able to choose the transition probabilities for each state from a given class and that our choice determines the cost incurred. The question arises as to how best to do this to minimize our expected costs.

Example 5.4.1

A random walker on $\{0, 1, 2, \dots\}$ jumps one step to the right with probability p and one step to the left with probability $q = 1 - p$. Any value of $p \in (0, 1]$ may be chosen, but incurs a cost

$$c(p) = 1/p.$$

The walker on reaching 0 stays there, incurring no further costs.

If we are only concerned with minimizing costs over the first few time steps, then the choice $p = 1$ may be best. However, in the long run the only way to avoid an infinite total cost is to get to 0. Starting from i we must first hit $i-1$, then $i-2$, and so on. Given the lack of memory in the model, this makes it reasonable to pick the same value of p throughout, and seek to minimize $\phi(p)$, the expected total cost starting from 1. The expected total cost starting from 2 is $2\phi(p)$ since we must first hit 1. Hence

$$\phi(p) = c(p) + 2p\phi(p)$$

so that

$$\phi(p) = \begin{cases} c(p)/(1-2p) & \text{for } p < 1/2 \\ \infty & \text{for } p \geq 1/2. \end{cases}$$

Thus for $c(p) = 1/p$ the choice $p = 1/4$ is optimal, with expected cost 8. The general discussion which follows will make rigorous what we claimed was reasonable.

Generally, let us suppose given some distribution $\lambda = (\lambda_i : i \in I)$ and, for each *action* $a \in A$, a transition matrix $P(a) = (p_{ij}(a) : i, j \in I)$ and a cost function $c(a) = (c_i(a) : i \in I)$. These are the data for a *Markov decision process*, though so far we have no process and when we do it will not in general be Markov. To get a process we must choose a policy, that is, a way of determining actions by our current knowledge of the process. Formally, a *policy* u is a sequence of functions

$$u_n : I^{n+1} \to A, \qquad n = 0, 1, 2, \ldots .$$

Each policy u determines a probability law $\mathbb{P}^u$ for a process $(X_n)_{n\geq 0}$ with values in I by

(i) $\mathbb{P}^u(X_0 = i_0) = \lambda_{i_0}$;
(ii) $\mathbb{P}^u(X_{n+1} = i_{n+1} \mid X_0 = i_0, \ldots, X_n = i_n) = p_{i_n i_{n+1}}(u_n(i_0, \ldots, i_n))$.

A *stationary policy* u is a function $u : I \to A$. We abuse notation and write u also for the associated policy given by

$$u_n(i_0, \ldots, i_n) = u(i_n).$$

Under a stationary policy u, the probability law $\mathbb{P}^u$ makes $(X_n)_{n\geq 0}$ Markov, with transition probabilities $p_{ij}^u = p_{ij}(u(i))$.

We suppose that a cost $c(i, a) = c_i(a)$ is incurred when action a is chosen in state i. Then we associate to a policy u an *expected total cost* starting from i, given by

$$V^u(i) = \mathbb{E}^u \sum_{n=0}^{\infty} c(X_n, u_n(X_0, \ldots, X_n)).$$

So that this sum is well defined, we assume that $c(i, a) \geq 0$ for all i and a. Define also the *value function*

$$V^*(i) = \inf_u V^u(i)$$

which is the minimal expected total cost starting from i.

The basic problem of Markov decision theory is how to minimize expected costs by our choice of policy. The minimum expected cost incurred before time $n = 1$ is given by

$$V_1(i) = \inf_a c(i, a).$$

Then the minimum expected cost incurred before time $n = 2$ is

$$V_2(i) = \inf_a \Big\{ c(i, a) + \sum_{j \in I} p_{ij}(a) V_1(j) \Big\}.$$

5.4 Markov decision processes

Define inductively
$$V_{n+1}(i) = \inf_a \left\{ c(i,a) + \sum_{j \in I} p_{ij}(a) V_n(j) \right\}. \tag{5.4}$$

It is easy to see by induction that $V_n(i) \leq V_{n+1}(i)$ for all i, so $V_n(i)$ increases to a limit $V_\infty(i)$, possibly infinite. We have
$$V_{n+1}(i) \leq c(i,a) + \sum_{j \in I} p_{ij}(a) V_n(j) \qquad \text{for all } a$$

so, letting $n \to \infty$ and then minimizing over a,
$$V_\infty(i) \leq \inf_a \left\{ c(i,a) + \sum_{j \in I} p_{ij}(a) V_\infty(j) \right\}. \tag{5.5}$$

It is a reasonable guess that $V_\infty(i)$, being the limit of minimal expected costs over finite time intervals, is in fact the value function $V^*(i)$. This is not always true, unless we can show that the inequality (5.5) is actually an equality. We make three technical assumptions to ensure this. *We assume that*

(i) *for all i,j the functions $c_i : A \to [0, \infty)$ and $p_{ij} : A \to [0, \infty)$ are continuous;*

(ii) *for all i and all $B < \infty$ the set $\{a : c_i(a) \leq B\}$ is compact;*

(iii) *for each i, for all but finitely many j, for all $a \in A$ we have $p_{ij}(a) = 0$.*

A simple case where (i) and (ii) hold is when A is a finite set. It is easy to check that the assumptions are valid in Example 5.4.1, with $A = (0,1]$, $c_i(a) = 1/a$ and
$$p_{ij}(a) = \begin{cases} a & \text{if } j = i+1 \\ 1-a & \text{if } j = i-1 \\ 0 & \text{otherwise,} \end{cases}$$

with obvious exceptions at $i = 0$.

Lemma 5.4.2. *There is a stationary policy u such that*
$$V_\infty(i) = c(i, u(i)) + \sum_{j \in I} p_{ij}(u(i)) V_\infty(j). \tag{5.6}$$

Proof. If $V_\infty(i) = \infty$ there is nothing to prove, so let us assume that $V_\infty(i) \leq B < \infty$. Then
$$V_{n+1}(i) = \inf_{a \in K} \left\{ c(i,a) + \sum_{j \in J} p_{ij}(a) V_n(j) \right\}$$

where K is the compact set $\{a : c(i,a) \leq B\}$ and where J is the finite set $\{j : p_{ij} \not\equiv 0\}$. Hence, by continuity, the infimum is attained and

$$V_{n+1}(i) = c(i, u_n(i)) + \sum_{j \in J} p_{ij}(u_n(i)) V_n(j) \qquad (5.7)$$

for some $u_n(i) \in K$. By compactness there is a convergent subsequence $u_{n_k}(i) \to u(i)$, say, and, on passing to the limit $n_k \to \infty$ in (5.7), we obtain (5.6). □

Theorem 5.4.3. *We have*
  (i) $V_n(i) \uparrow V^*(i)$ *as* $n \to \infty$ *for all* i;
  (ii) *if* u^* *is any stationary policy such that* $a = u^*(i)$ *minimizes*

$$c(i, a) + \sum_{j \in I} p_{ij}(a) V^*(j)$$

for all i, *then* u^* *is* optimal, *in the sense that*

$$V^{u^*}(i) = V^*(i) \qquad \text{for all } i.$$

Proof. For any policy u we have

$$V^u(i) = \mathbb{E}_i^u \sum_{n=0}^{\infty} c(X_n, u_n(X_0, \ldots, X_n))$$
$$= c(i, u_0(i)) + \sum_{j \in I} p_{ij}(u_0(i)) V^{u[i]}(j)$$

where $u[i]$ is the policy given by

$$u[i]_n(i_0, \ldots, i_n) = u_{n+1}(i, i_0, \ldots, i_n).$$

Hence we obtain

$$V^u(i) \geq \inf_a \Big\{ c(i, a) + \sum_{j \in I} p_{ij}(a) V^*(j) \Big\}$$

and, on taking the infimum over u

$$V^*(i) \geq \inf_a \Big\{ c(i, a) + \sum_{j \in I} p_{ij}(a) V^*(j) \Big\}. \qquad (5.8)$$

Certainly, $V_0(i) = 0 \leq V^*(i)$. Let us suppose inductively that $V_n(i) \leq V^*(i)$ for all i. Then by substitution in the right sides of (5.4) and (5.8) we find $V_{n+1}(i) \leq V^*(i)$ and the induction proceeds. Hence $V_\infty(i) \leq V^*(i)$ for all i.

5.4 Markov decision processes

Let u^* be any stationary policy for which
$$V_\infty(i) \geq c(i, u^*(i)) + \sum_{j \in I} p_{ij}(u^*(i)) V_\infty(j).$$

We know such a policy exists by Lemma 5.4.2. Then by Theorem 4.2.3 we have $V^{u^*}(i) \leq V_\infty(i)$ for all i. But $V^*(i) \leq V^{u^*}(i)$ for all i, so
$$V_\infty(i) = V^*(i) = V^{u^*}(i) \qquad \text{for all } i$$
and we are done. □

The theorem just proved shows that the problem of finding a good policy is much simpler than we might have supposed. For it was not clear at the outset that there would be a single policy which was optimal for all i, even less that this policy would be stationary. Moreover, part (i) gives an explicit way of obtaining the value function V^* and, once this is known, part (ii) identifies an optimal stationary policy.

In practice we may know only an approximation to V^*, for example V_n for n large. We may then hope that, by choosing $a = u(i)$ to minimize
$$c(i, a) + \sum_{j \in I} p_{ij}(a) V_n(j)$$

we get a nearly optimal policy. An alternative means of constructing nearly optimal policies is sometimes provided by the method of *policy improvement*. Given one stationary policy u we may define another θu by the requirement that $a = (\theta u)(i)$ minimizes
$$c(i, a) + \sum_{j \in I} p_{ij}(a) V^u(j).$$

Theorem 5.4.4 (Policy improvement). *We have*
(i) $V^{\theta u}(i) \leq V^u(i)$ *for all* i;
(ii) $V^{\theta^n u}(i) \downarrow V^*(i)$ *as* $n \to \infty$ *for all* i, *provided that*
$$\mathbb{E}_i^{u^*}(V^u(X_n)) \to 0 \quad \text{as} \quad n \to \infty \qquad \text{for all } i. \tag{5.9}$$

Proof. (i) We have, by Theorem 4.2.3,
$$V^u(i) = c(i, u(i)) + \sum_{j \in I} p_{ij}(u(i)) V^u(j)$$
$$\geq c(i, \theta u(i)) + \sum_{j \in I} p_{ij}(\theta u(i)) V^u(j)$$

so $V^u(i) \geq V^{\theta u}(i)$ for all i, by Theorem 4.2.3.

(ii) We note from part (i) that

$$V^{\theta u}(i) \leq c(i,a) + \sum_{j \in I} p_{ij}(a) V^u(j) \qquad \text{for all } i \text{ and } a. \qquad (5.10)$$

Fix $N \geq 0$ and consider for $n = 0, 1, \ldots, N$ the process

$$M_n = V^{\theta^{N-n} u}(X_n) + \sum_{k=0}^{n-1} c(X_k, u^*(X_k)).$$

Recall the notation for conditional expectation introduced in Section 4.1. We have

$$\mathbb{E}^{u^*}(M_{n+1} \mid \mathcal{F}_n) = \sum_{j \in I} p_{X_n j}(u^*(X_n)) V^{\theta^{N-n-1} u}(j) + c(X_n, u^*(X_n))$$

$$+ \sum_{k=0}^{n-1} c(X_k, u^*(X_k))$$

$$\geq M_n$$

where we used (5.10) with u replaced by $\theta^{N-n-1} u$, $i = X_n$ and $a = u^*(X_n)$. It follows that $\mathbb{E}^{u^*}(M_{n+1}) \geq \mathbb{E}^{u^*}(M_n)$ for all n. Hence if we assume (5.9), then

$$V^{\theta^N u}(i) = \mathbb{E}_i^{u^*}(M_0) \leq \mathbb{E}_i^{u^*}(M_N)$$

$$= \mathbb{E}_i^{u^*}(V^u(X_N)) + \mathbb{E}^{u^*}\left(\sum_{n=0}^{N-1} c(X_n, u^*(X_n))\right)$$

$$\to V^*(i) \qquad \text{as } N \to \infty. \qquad \square$$

We have been discussing the minimization of expected total cost, which is only relevant to the transient case. This is because we will have $V^*(i) = \infty$ unless for some stationary policy u, the only states j with positive cost $c(j, u(j)) > 0$, accessible from i, are transient. The recurrent case is also of practical importance and one way to deal with this is to discount costs at future times by a fixed factor $\alpha \in (0,1)$. We now seek to minimize the *expected total discounted cost*

$$V_\alpha^u(i) = \mathbb{E}_i^u \sum_{n=0}^\infty \alpha^n c(X_n, u_n(X_0, \ldots, X_n)).$$

Define the *discounted value function*

$$V_\alpha^*(i) = \inf_u V_\alpha^u(i).$$

In fact, the discounted case reduces to the undiscounted case by introducing a new absorbing state ∂ and defining a new Markov decision process by
$$\widetilde{p}_{ij}(a) = \alpha p_{ij}(a), \quad \widetilde{p}_{i\partial}(a) = 1 - \alpha,$$
$$\widetilde{c}_i(a) = c_i(a), \quad \widetilde{c}_\partial(a) = 0.$$

Thus the new process follows the old until, at some geometric time of parameter α, it jumps to ∂ and stays there, incurring no further costs.

Introduce $V_{0,\alpha}(i) = 0$ and, inductively
$$V_{n+1,\alpha}(i) = \inf_a \Big\{ c(i,a) + \alpha \sum_{j \in J} p_{ij}(a) V_{n,\alpha}(i) \Big\}$$

and, given a stationary policy u, define another $\theta_\alpha u$ by the requirement that $a = (\theta_\alpha u)(i)$ minimizes
$$c(i,a) + \alpha \sum_{j \in J} p_{ij}(a) V^u(j).$$

Theorem 5.4.5. *Suppose that the cost function $c(i,a)$ is uniformly bounded.*

(i) *We have $V_{n,\alpha}(i) \uparrow V^*_\alpha(i)$ as $n \to \infty$ for all i.*

(ii) *The value function V^*_α is the unique bounded solution to*
$$V^*_\alpha(i) = \inf_a \Big\{ c(i,a) + \alpha \sum_{j \in I} p_{ij}(a) V^*_\alpha(j) \Big\}. \tag{5.11}$$

(iii) *Let u^* be a stationary policy such that $a = u^*(i)$ minimizes*
$$c(i,a) + \alpha \sum_{j \in I} p_{ij}(a) V^*_\alpha(j)$$
for all i. Then u^* is optimal in the sense that
$$V^{u^*}_\alpha(i) = V^*_\alpha(i) \qquad \text{for all } i.$$

(iv) *For all stationary policies u we have*
$$V^{\theta^n_\alpha u}_\alpha(i) \downarrow V^*_\alpha(i) \qquad \text{as } n \to \infty \text{ for all } i.$$

Proof. With obvious notation we have
$$V^u_\alpha = \widetilde{V}^u, \quad V^*_\alpha = \widetilde{V}^*, \quad V_{n,\alpha} = \widetilde{V}_n, \quad \theta_\alpha u = \widetilde{\theta}_u.$$

so parts (i), (ii) and (iii) follow directly from Theorems 5.4.3 and 5.4.4, except for the uniqueness claim in (ii). But given any bounded solution V to (5.11), there is a stationary policy u such that

$$V(i) = c(i, u(i)) + \alpha \sum_{j \in I} p_{ij}(u(i)) V(j).$$

Then $V = V_\alpha^u$, by Theorem 4.2.5. Then $\theta_\alpha u = u$ so (iv) will show that u is optimal and $V = V_\alpha^*$.

We have $c(i, a) \leq B$ for some $B < \infty$. So for any stationary policy u we have

$$V_\alpha^u(i) = \mathbb{E}_i^u \sum_{n=0}^{\infty} \alpha^n c(X_n, u(X_n)) \leq B/(1-\alpha)$$

and so

$$\widetilde{\mathbb{E}}_i^{u^*}\big(\widetilde{V}^u(X_n)\big) = \alpha^n \mathbb{E}_i^{u^*}\big(V_\alpha^u(X_n)\big) \leq B\alpha^n/(1-\alpha) \to 0$$

as $n \to \infty$. Hence (iv) also follows from Theorem 5.4.4. □

We finish with a discussion of long-run average costs. Here we are concerned with the limiting behaviour, as $n \to \infty$, of

$$\overline{V}_n^u(i) = \mathbb{E}_i^u \left(\frac{1}{n} \sum_{k=0}^{n-1} c(X_k, u_k(X_0, \ldots, X_k)) \right).$$

We assume that

$$|c(i, a)| \leq B < \infty \qquad \text{for all } i \text{ and } a.$$

This forces $|\overline{V}_n^u(i)| \leq B$ for all n, but in general the sequence $\overline{V}_n^u(i)$ may fail to converge as $n \to \infty$. In the case of a stationary strategy u for which $(X_n)_{n \geq 0}$ has a unique invariant distribution π^u, we know by the ergodic theorem that

$$\frac{1}{n} \sum_{k=0}^{n-1} c(X_k, u(X_k)) \to \sum_{j \in I} \pi_j^u c(j, u(j))$$

as $n \to \infty$, $\mathbb{P}_i^u$-almost surely, for all i. So $\overline{V}_n^u(i)$ does converge in this case by bounded convergence, with the same limit. This suggests that one approach to minimizing long-run costs might be to minimize

$$\sum_{j \in I} \pi_j^u c(j, u(j)).$$

But, although this is sometimes valid, we do not know in general that the optimal policy is positive recurrent, or even stationary. Instead, we use a martingale approach, which is more general.

Theorem 5.4.6. Suppose we can find a constant $\overline{V}^*$ and a bounded function $W(i)$ such that
$$\overline{V}^* + W(i) = \inf_a \Big\{ c(i,a) + \sum_{j \in I} p_{ij}(a) W(j) \Big\} \qquad \text{for all } i. \tag{5.12}$$

Let u^* be any stationary strategy such that $a = u^*(i)$ achieves the infimum in (5.12) for each i. Then

(i) $\overline{V}_n^{u^*}(i) \to \overline{V}^*$ as $n \to \infty$ for all i;
(ii) $\liminf_{n \to \infty} \overline{V}_n^u(i) \geq \overline{V}^*$ for all i, for all u.

Proof. Fix a strategy u and set $U_n = u_n(X_0, \ldots, X_n)$. Consider
$$M_n = W(X_n) - n\overline{V}^* + \sum_{k=0}^{n-1} c(X_k, U_k).$$

Then
$$\mathbb{E}^u(M_{n+1} \mid \mathcal{F}_n)$$
$$= M_n + \Big\{ c(X_n, U_n) + \sum_{j \in I} p_{X_n j}(U_n) W(j) \Big\} - \big(\overline{V}^* + W(X_n)\big)$$
$$\geq M_n$$

with equality if $u = u^*$. Therefore
$$W(i) = \mathbb{E}_i^u(M_0) \leq \mathbb{E}_i^u(M_n) = \mathbb{E}_i^u W(X_n) - n\overline{V}^* + n\overline{V}_n^u(i).$$

So we obtain
$$\overline{V}^* \leq \overline{V}_n^u(i) + 2 \sup_i |W(i)|/n.$$

This implies (ii) on letting $n \to \infty$. When $u = u^*$ we also have
$$\overline{V}_n^{u^*}(i) \leq \overline{V}^* + 2 \sup_i |W(i)|/n$$

and hence (i). $\square$

The most obvious point of this theorem is that it identifies an optimal stationary policy when the hypothesis is met. Two further aspects also deserve comment. Firstly, if u is a stationary policy for which $(X_n)_{n \geq 0}$ has an invariant distribution π^u, then

$$\sum_{i \in I} \pi_i^u \big(\overline{V}^* + W(i)\big) \leq \sum_{i \in I} \pi_i^u \bigg(c\big(i, u(i)\big) + \sum_{j \in I} p_{ij}\big(u(i)\big) W(j) \bigg)$$
$$= \sum_{i \in I} \pi_i^u c\big(i, u(i)\big) + \sum_{j \in I} \pi_j^u W(j)$$

so

$$\overline{V}^* \leq \sum_{i \in I} \pi_i^u c(i, u(i))$$

with equality if we can take $u = u^*$.

Secondly, there is a connection with the case of discounted costs. Assume that I is finite and that $P(a)$ is irreducible for all a. Then we can show that as $\alpha \uparrow 1$ we have

$$V_\alpha^*(i) = \overline{V}^*/(1-\alpha) + W(i) + o(1-\alpha).$$

On substituting this into (5.11) we find

$$\overline{V}^*/(1-\alpha) + W(i) + o(1-\alpha)$$
$$= \inf_a \left\{ c(i,a) + \alpha \sum_{j \in I} p_{ij}(a)(\overline{V}^*/(1-\alpha) + W(j) + o(1-\alpha)) \right\}$$

so

$$\overline{V}^* + W(i) = \inf_a \left\{ c(i,a) + \alpha \sum_{j \in I} p_{ij}(a) W(j) \right\} + o(1-\alpha)$$

which brings us back to (5.12) on letting $\alpha \uparrow 1$.

The interested reader is referred to S. M. Ross, *Applied Probability Models with Optimization Applications* (Holden-Day, San Francisco, 1970) and to H. C. Tijms, *Stochastic Models – an algorithmic approach* (Wiley, Chichester, 1994) for more examples, results and references.

5.5 Markov chain Monte Carlo

Most computers may be instructed to provide a sequence of numbers

$$u_1 = 0.u_{11}u_{12}u_{13}\ldots u_{1m}$$
$$u_2 = 0.u_{21}u_{22}u_{23}\ldots u_{2m}$$
$$u_3 = 0.u_{31}u_{32}u_{33}\ldots u_{3m}$$

written as decimal expansions of a certain length, which for many purposes may be regarded as sample values of a sequence of independent random variables, uniformly distributed on $[0,1]$:

$$U_1(\omega), U_2(\omega), U_3(\omega), \ldots.$$

5.5 Markov chain Monte Carlo

We are cautious in our language because, of course, $u_1, u_2, u_3, \ldots$ are actually all integer multiples of 10^{-m} and, more seriously, they are usually derived sequentially by some entirely deterministic algorithm in the computer. Nevertheless, the generators of such *pseudo-random numbers* are in general as reliable an imitation as one could wish of $U_1(\omega), U_2(\omega), U_3(\omega), \ldots$. This makes it worth while considering how one might construct Markov chains from a given sequence of independent uniform random variables, and then might exploit the observed properties of such processes.

We shall now describe one procedure to simulate a Markov chain $(X_n)_{n \geq 0}$ with initial distribution λ and transition matrix P. Since $\sum_{i \in I} \lambda_i = 1$ we can partition $[0, 1]$ into disjoint subintervals $(A_i : i \in I)$ with lengths

$$|A_i| = \lambda_i.$$

Similarly for each $i \in I$, we can partition $[0, 1]$ into disjoint subintervals $(A_{ij} : j \in I)$ such that

$$|A_{ij}| = p_{ij}.$$

Now define functions

$$G_0 : [0,1] \to I,$$
$$G : I \times [0,1] \to I$$

by

$$G_0(u) = i \quad \text{if } u \in A_i,$$
$$G(i, u) = j \quad \text{if } u \in A_{ij}.$$

Suppose that $U_0, U_1, U_2, \ldots$ is a sequence of independent random variables, uniformly distributed on $[0, 1]$, and set

$$X_0 = G_0(U_0),$$
$$X_{n+1} = G(X_n, U_{n+1}) \quad \text{for } n \geq 0.$$

Then

$$\mathbb{P}(X_0 = i) = \mathbb{P}(U_0 \in A_i) = \lambda_i,$$
$$\mathbb{P}(X_{n+1} = i_{n+1} \mid X_0 = i_0, \ldots, X_n = i_n) = \mathbb{P}(U_{n+1} \in A_{i_n i_{n+1}}) = p_{i_n i_{n+1}}$$

so $(X_n)_{n \geq 0}$ is Markov(λ, P).

This simple procedure may be used to investigate empirically those aspects of the behaviour of a Markov chain where theoretical calculations become infeasible.

The remainder of this section is devoted to one application of the simulation of Markov chains. It is the application which finds greatest practical use, especially in statistics, statistical physics and computer science, known as *Markov chain Monte Carlo*. Monte Carlo is another name for computer simulation so this sounds no different from the procedure just discussed. But what is really meant is simulation *by means of* Markov chains, the object of primary interest being the invariant distribution of the Markov chain and not the chain itself. After a general discussion we shall give two examples.

The context for Markov chain Monte Carlo is a state-space in product form
$$I = \prod_{m \in \Lambda} S_m$$
where Λ is a finite set. For the purposes of this discussion we shall also assume that each component S_m is a finite set. A random variable X with values in I is then a family of component random variables $(X(m) : m \in \Lambda)$, where, for each site $m \in \Lambda$, $X(m)$ takes values in S_m.

We are given a distribution $\pi = (\pi_i : i \in I)$, perhaps up to an unknown constant multiple, and it is desired to compute the number
$$\sum_{i \in I} \pi_i f_i \qquad (5.13)$$
for some given function $f = (f_i : i \in I)$. The essential point to understand is that Λ is typically a large set, making the state-space I very large indeed. Then certain operations are computationally infeasible – performing the sum (5.13) state by state for a start.

An alternative approach would be to simulate a large number of independent random variables $X_1, \ldots, X_n$ in I, each with distribution π, and to approximate (5.13) by
$$\frac{1}{n} \sum_{k=1}^{n} f(X_k).$$
The strong law of large numbers guarantees that this is a good approximation as $n \to \infty$ and, moreover, one can obtain error estimates which indicate how large to make n in practice. However, simulation from the distribution π is also difficult, unless π has product form
$$\pi(x) = \prod_{m \in \Lambda} \pi_m(x(m)).$$

For recall that a computer just simulates sequences of independent $U[0,1]$ random variables. When π does not have product form, Markov chain Monte Carlo is sometimes the only way to simulate samples from π.

5.5 Markov chain Monte Carlo

The basic idea is to simulate a Markov chain $(X_n)_{n\geq 0}$, which is constructed to have invariant distribution π. Then, assuming aperiodicity and irreducibility, we know, by Theorem 1.8.3, that as $n \to \infty$ the distribution of X_n converges to π. Indeed, assuming only irreducibility, Theorem 1.10.2 shows that

$$\frac{1}{n}\sum_{k=0}^{n-1} f(X_k) \to \sum_{i\in I} \pi_i f_i$$

with probability 1. But why should simulating an entire Markov chain be easier than simulating a simple distribution π? The answer lies in the fact that the state-space is a product.

Each component $X_0(m)$ of the initial state X_0 is a random variable in S_m. It does not matter crucially what distribution X_0 is given, but we might, for example, make all components independent. The process $(X_n)_{n\geq 0}$ is made to evolve by changing components one site at a time. When the chosen site is m, we simulate a new random variable $X_{n+1}(m)$ with values in S_m according to a distribution determined by X_n, and for $k \neq m$ we set $X_{n+1}(k) = X_n(k)$. Thus at each step we have only to simulate a random variable in S_m, not one in the much larger space I.

Let us write $i \overset{m}{\sim} j$ if i and j agree, except possibly at site m. The law for simulating a new value at site m is described by a transition matrix $P(m)$, where

$$p_{ij}(m) = 0 \quad \text{unless } i \overset{m}{\sim} j.$$

We would like π to be invariant for $P(m)$. A sufficient condition is that the detailed balance equations hold: thus for all i,j we want

$$\pi_i p_{ij}(m) = \pi_j p_{ji}(m).$$

There are many possible choices for $P(m)$ satisfying these equations. Indeed, given any stochastic matrix $R(m)$ with

$$r_{ij}(m) = 0 \quad \text{unless } i \overset{m}{\sim} j$$

we can determine such a $P(m)$ by

$$\pi_i p_{ij}(m) = \big(\pi_i r_{ij}(m)\big) \wedge \big(\pi_j r_{ji}(m)\big)$$

for $i \neq j$, and then

$$p_{ii}(m) = 1 - \sum_{j\neq i} p_{ij}(m) \geq 0.$$

This has the following interpretation: if $X_n = i$ we simulate a new random variable Y_n so that $Y_n = j$ with probability $r_{ij}(m)$, then if $Y_n = j$ we set

$$X_{n+1} = \begin{cases} Y_n & \text{with probability } \bigl(\pi_j r_{ji}(m)/\pi_i r_{ij}(m)\bigr) \wedge 1 \\ X_n & \text{otherwise.} \end{cases}$$

This is called a *Hastings algorithm*.

There are two commonly used special cases. On taking

$$r_{ij}(m) = \Bigl(\sum_{k \stackrel{m}{\sim} i} \pi_k\Bigr)^{-1} \pi_j \qquad \text{for } i \stackrel{m}{\sim} j$$

we also find

$$p_{ij}(m) = \Bigl(\sum_{k \stackrel{m}{\sim} i} \pi_k\Bigr)^{-1} \pi_j \qquad \text{for } i \stackrel{m}{\sim} j.$$

So we simply resample $X_n(m)$ according to the conditional distribution under π, given the other components. This is called the *Gibbs sampler*. It is particularly useful in Bayesian statistics.

On taking $r_{ij}(m) = r_{ji}(m)$ for all i and j we find

$$p_{ij}(m) = \bigl((\pi_j/\pi_i) \wedge 1\bigr) r_{ij}(m) \qquad \text{for } i \stackrel{m}{\sim} j, i \neq j.$$

This is called a *Metropolis algorithm*. A particularly simple case would be to take

$$r_{ij}(m) = 1/(N_m - 1) \qquad \text{for } i \stackrel{m}{\sim} j, i \neq j$$

where $N_m = |S_m|$. This amounts to choosing another value j_m at site m uniformly at random; if $\pi_j > \pi_i$, then we adopt the new value, whereas if $\pi_j \leq \pi_i$ we adopt the new value with probability π_j/π_i.

We have not yet specified a rule for deciding which site to visit when. In practice this may not matter much, provided we keep returning to every site. For definiteness we mention two possibilities. We might choose to visit every site once and then repeat, generating a sequence of sites $(m_n)_{n \geq 0}$. Then $(m_n, X_n)_{n \geq 0}$ is a Markov chain in $\Lambda \times I$. Alternatively, we might choose a site randomly at each step. Then $(X_n)_{n \geq 0}$ is itself a Markov chain with transition matrix

$$P = |\Lambda|^{-1} \sum_{m \in \Lambda} P(m).$$

We shall stick with this second choice, where the analysis is simpler to present. Let us assume that P is irreducible, which is easy to ensure in the examples. We know that

$$\pi_i p_{ij}(m) = \pi_j p_{ji}(m)$$

for all m and all i, j, so also
$$\pi_i p_{ij} = \pi_j p_{ji}$$
and so π is the unique invariant measure for P. Hence, by Theorem 1.10.2, we have
$$\frac{1}{n} \sum_{k=0}^{n-1} f(X_k) \to \sum_{i \in I} \pi_i f_i$$
as $n \to \infty$ with probability 1. Thus the algorithm works eventually. In practice one is concerned with how fast it works, but useful information of this type cannot be gained in the present general context. Given more information on the structure of S_m and the distribution π to be simulated, much more can be said. We shall not pursue the matter here. It should also be emphasised that there is an empirical side to simulation: with due caution informed by the theory, the computer output gives a good idea of how well we are doing. For further reading we recommend *Stochastic Simulation* by B. D. Ripley (Wiley, Chichester, 1987), and *Markov Chain Monte Carlo in practice* by W. R. Gilks, S. Richardson and D. J. Spiegelhalter (Chapman and Hall, London, 1996). The recent survey article *Bayesian computation and stochastic systems* by J. Besag, P. Green, D. Higdon and K. Mengersen (*Statistical Science*, 10 (1), pp. 3–40, 1995) contains many interesting references. We finish with two examples.

Example 5.5.1 (Bayesian statistics)

In a statistical problem one may be presented with a set of independent observations $Y_1, \ldots, Y_n$, which it is reasonable to assume are normally distributed, but with unknown mean μ and variance τ^{-1}. One then seeks to draw conclusions about μ and τ on the basis of the observations. The Bayesian approach to this problem is to assume that μ and τ are themselves random variables, with a given prior distribution. For example, we might assume that
$$\mu \sim N(\theta_0, \phi_0^{-1}), \quad \tau \sim \Gamma(\alpha_0, \beta_0),$$
that is to say, μ is normal of mean θ_0 and variance ϕ_0^{-1}, and τ has gamma distribution of parameters α_0 and β_0. The parameters θ_0, ϕ_0, α_0 and β_0 are known. Then the prior density for (μ, τ) is given by
$$\pi(\mu, \tau) \propto \exp\{-\phi_0(\mu - \theta_0)^2/2\} \tau^{\alpha_0 - 1} \exp\{-\beta_0 \tau\}.$$

The posterior density for (μ, τ), which is the conditional density given the observations, is then given by Bayes' formula
$$\pi(\mu, \tau \mid y) \propto \pi(\mu, \tau) f(y \mid \mu, \tau)$$
$$\propto \exp\{-\phi_0(\mu - \theta_0)^2/2\} \exp\left\{-\tau \sum_{i=1}^n (y_i - \mu)^2/2\right\} \tau^{\alpha_0 - 1 + n/2} \exp\{-\beta_0 \tau\}.$$

Note that the posterior density is no longer in product form: the conditioning has introduced a dependence between μ and τ. Nevertheless, the *full conditional distributions* still have a simple form

$$\pi(\mu \mid y, \tau) \propto \exp\{-\phi_0(\mu - \theta_0)^2/2\} \exp\left\{-\tau \sum_{i=1}^n (y_i - \mu)^2/2\right\} \sim N(\theta_n, \phi_n^{-1}),$$

$$\pi(\tau \mid y, \mu) \propto \tau^{\alpha_0 - 1 + n/2} \exp\left\{-\tau\left(\beta_0 + \sum_{i=1}^n (y_i - \mu)^2/2\right)\right\} \sim \Gamma(\alpha_n, \beta_n)$$

where

$$\theta_n = \left(\phi_0 \theta_0 + \tau \sum_{i=1}^n y_i\right)/(\phi_0 + n\tau), \quad \phi_n = \phi_0 + n\tau,$$

$$\alpha_n = \alpha_0 + n/2, \quad \beta_n = \beta_0 + \sum_{i=1}^n (y_i - \mu)^2/2.$$

Our final belief about μ and τ is regarded as measured by the posterior density. We may wish to compute probabilities and expectations. Here the *Gibbs sampler* provides a particularly simple approach. Of course, numerical integration would also be feasible as the dimension is only two. To make the connection with our general discussion we set

$$I = S_1 \times S_2 = \mathbb{R} \times [0, \infty).$$

We wish to simulate $X = (\mu, \tau)$ with density $\pi(\mu, \tau \mid y)$. The fact that $\mathbb{R}$ and $[0, \infty)$ are not finite sets does not affect the basic idea. In any case the computer will work with finite approximations to $\mathbb{R}$ and $[0, \infty)$. First we simulate X_0, say from the product form density $\pi(\mu, \tau)$. At the kth stage, given $X_k = (\mu_k, \tau_k)$, we first simulate μ_{k+1} from $\pi(\mu \mid y, \tau_k)$ and then τ_{k+1} from $\pi(\tau \mid y, \mu_{k+1})$, then set $X_{k+1} = (\mu_{k+1}, \tau_{k+1})$. Then $(X_k)_{k \geq 0}$ is a Markov chain in I with invariant measure $\pi(\mu, \tau \mid y)$, and one can show that

$$\frac{1}{k} \sum_{j=0}^{k-1} f(X_j) \to \int_I f(x) \pi(x \mid y) dx \quad \text{as } k \to \infty$$

with probability 1, for all bounded continuous functions $f : I \to \mathbb{R}$. This is not an immediate consequence of the ergodic theorem for discrete state-space, but you may find it reasonable at an intuitive level, with a rate of convergence depending on the smoothness of π and f.

We now turn to an elaboration of this example where the Gibbs sampler is indispensible. The model consists of m copies of the preceding one, with different means but a common variance. Thus there are mn independent observations Y_{ij}, where $i = 1, \ldots n$, and $j = 1, \ldots, m$, normally distributed, with means μ_j and common variance τ^{-1}. We take these parameters to be independent random variables as before, with

$$\mu_j \sim N(\theta_0, \phi_0^{-1}), \quad \tau \sim \Gamma(\alpha_0, \beta_0).$$

Let us write $\mu = (\mu_1, \ldots, \mu_n)$. The prior density is given by

$$\pi(\mu, \tau) \propto \exp\left\{-\phi_0 \sum_{j=1}^m (\mu_j - \theta_0)^2/2\right\} \tau^{\alpha_0 - 1} \exp\{-\beta_0 \tau\}$$

and the posterior density is given by

$$\pi(\mu, \tau \mid y) \propto \exp\left\{-\phi_0 \sum_{j=1}^m (\mu_j - \theta_0)^2/2\right\}$$

$$\times \exp\left\{-\tau \sum_{i=1}^n \sum_{j=1}^m (y_{ij} - \mu_j)^2/2\right\} \tau^{\alpha_0 - 1 + mn/2} \exp\{-\beta_0 \tau\}.$$

Hence the full conditional distributions are

$$\pi(\mu_j \mid y, \tau) \sim N(\theta_{jn}, \phi_n^{-1}), \quad \pi(\tau \mid y, \mu) \sim \Gamma(\alpha_n, \beta_n)$$

where

$$\theta_{jn} = \left(\phi_0 \theta_0 + \tau \sum_{i=1}^n y_{ij}\right)/(\phi_0 + n\tau), \quad \phi_n = \phi_0 + n\tau,$$

$$\alpha_n = \alpha_0 + mn/2, \quad \beta_n = \beta_0 + \sum_{i=1}^n \sum_{j=1}^m (y_{ij} - \mu_j)^2/2.$$

We can construct approximate samples from $\pi(\mu, \tau \mid y)$, just as in the case $m = 1$ discussed above, by a Gibbs sampler method. Note that, conditional on τ, the means μ_j, for $j = 1, \ldots, m$, remain independent. Thus one can update all the means simultaneously in the Gibbs sampler. This has the effect of speeding convergence to the equilibrium distribution. In cases where m is large, numerical integration of $\pi(\mu, \tau \mid y)$ is infeasible, as is direct simulation from the distribution, so the Markov chain approach is the only one available.

Example 5.5.2 (Ising model and image analysis)

Consider a large box $\Lambda = \Lambda_N$ in $\mathbb{Z}^2$

$$\Lambda = \{-N, \ldots, -1, 0, 1, \ldots, N\}^2$$

with boundary $\partial \Lambda = \Lambda_N \setminus \Lambda_{N-1}$, and the *configuration space*

$$I = \{-1, 1\}^\Lambda.$$

For $x \in \Lambda$ define
$$H(x) = \tfrac{1}{2} \sum \bigl(x(m) - x(m')\bigr)^2$$

where the sum is taken over all pairs $\{m, m'\} \subseteq \Lambda$ with $|m - m'| = 1$. Note that $H(x)$ is small when the values taken by x at neighbouring sites are predominantly the same. We write

$$I^+ = \{x \in I : x(m) = 1 \quad \text{for all } m \in \partial \Lambda\}$$

and for each $\beta > 0$ define a probability distribution $\bigl(\pi(x) : x \in I^+\bigr)$ by

$$\pi(x) \propto e^{-\beta H(x)}.$$

As $\beta \downarrow 0$ the weighting becomes uniform, whereas, as $\beta \uparrow \infty$ the mass concentrates on configurations x where $H(x)$ is small. This is one of the fundamental models of statistical physics, called the *Ising model*. A famous and deep result of Onsager says that if X has distribution π, then

$$\lim_{N \to \infty} \mathbb{E}\bigl(X(0)\bigr) = \bigl[\bigl(1 - (\sinh 2\beta)^{-4}\bigr)^+\bigr]^{1/8}.$$

In particular, if $\sinh 2\beta \leq 1$, the fact that X is forced to take boundary values 1 does not significantly affect the distribution of $X(0)$ when N is large, whereas if $\sinh 2\beta > 1$ there is a residual effect of the boundary values on $X(0)$, uniformly in N.

Here we consider the problem of simulating the Ising model. Simulations may sometimes be used to guide further developments in the theory, or even to detect phenomena quite out of reach of the current theory. In fact, the Ising model is rather well understood theoretically; but there are many related models which are not, where simulation is still possible by simple modifications of the methods presented here.

First we describe a Gibbs sampler. Consider the sets of even and odd sites

$$\Lambda^+ = \{(m_1, m_2) \in \Lambda : m_1 + m_2 \text{ is even}\},$$
$$\Lambda^- = \{(m_1, m_2) \in \Lambda : m_1 + m_2 \text{ is odd}\}$$

and for $x \in I$ set
$$x^{\pm} = \bigl(x(m) : m \in \Lambda^{\pm}\bigr).$$

We can exploit the fact that the conditional distribution $\pi(x^+ \mid x^-)$ has product form
$$\pi(x^+ \mid x^-) \propto \prod_{m \in \Lambda^+ \setminus \partial \Lambda} e^{\beta x(m) s(m)}$$

where, for $m \in \Lambda^+ \setminus \partial \Lambda$
$$s(m) = \sum_{|m'-m|=1} x^-(m').$$

Therefore, it is easy to simulate from $\pi(x^+ \mid x^-)$ and likewise from $\pi(x^- \mid x^+)$. Choose now some simple initial configuration X_0 in I^+. Then inductively, given $X_n^- = x^-$, simulate firstly X_{n+1}^+ with distribution $\pi(\cdot \mid x^-)$ and then given $X_{n+1}^+ = x^+$, simulate X_{n+1}^- with distribution $\pi(\cdot \mid x^+)$. Then according to our general discussion, for large n, the distribution of X_n is approximately π. Note that we did not use the value of the normalizing constant
$$Z = \sum_{x \in I^+} e^{-\beta H(x)}$$

which is hard to compute by elementary means when N is large.

An alternative approach is to use a Metropolis algorithm. We can again exploit the even/odd partition. Given that $X_n = x$, independently for each $m \in \Lambda^+ \setminus \partial \Lambda$, we change the sign of $X_n^+(m)$ with probability
$$p(m,x) = \bigl(\pi(\hat{x})/\pi(x)\bigr) \wedge 1 = e^{2\beta x(m) s(m)} \wedge 1$$

where $\hat{x} \overset{m}{\sim} x$ with $\hat{x}(m) = -x(m)$. Let us call the resulting configuration Y_n. Next we apply the corresponding transformation to $Y_n^-(m)$ for the odd sites $m \in \Lambda^- \setminus \partial \Lambda$, to obtain X_{n+1}. The process $(X_n)_{n \geq 0}$ is then a Markov chain in I^+ with invariant distribution π.

Both methods we have described serve to simulate samples from π; there is little to choose between them. Convergence is fast in the subcritical case $\sinh 2\beta < 1$, where π has an approximate product structure on large scales.

In a Bayesian analysis of two-dimensional images, the Ising model is sometimes used as a prior. We may encode a digitized image on a two-dimensional grid as a particular configuration $\bigl(x(m) : m \in \Lambda\bigr) \in I$, where $x(m) = 1$ for a white pixel and $x(m) = -1$ for a black pixel. By varying the parameter β in the Ising model, we vary the tendency of black pixels to clump together; the same for white pixels. Thus β is a sort of texture parameter, which we choose according to the sort of image we expect, thus obtaining a prior $\pi(x)$. Observations are now made at each site which record the true pixel, black or white, with probability $p \in (0,1)$. The posterior distribution for X given observations Y is then given by

$$\pi(x \mid y) \propto \pi(x) f(y \mid x) \propto e^{-\beta H(x)} p^{a(x,y)} (1-p)^{d(x,y)}$$

where $a(x,y)$ and $d(x,y)$ are the numbers of sites at which x and y agree and disagree respectively. 'Cleaned-up' versions of the observed image Y may now be obtained by simulating from the posterior distribution. Although this is not exactly the Ising model, the same methods work. We describe the appropriate Metropolis algorithm: given that $X_n = x$, independently for each $m \in \Lambda^+ \setminus \partial \Lambda$, change the sign of $X_n^+(m)$ with probability

$$\begin{aligned} p(m,x,y) &= \big(\pi(\hat{x} \mid y)/\pi(x \mid y)\big) \wedge 1 \\ &= e^{-2\beta x(m) s(m)} \big((1-p)/p\big)^{x(m) y(m)} \end{aligned}$$

where $\hat{x} \stackrel{m}{\sim} x$ with $\hat{x}(m) = -x(m)$. Call the resulting configuration $X_{n+1/2}$. Next apply the corresponding transformation to $X_{n+1/2}^-$ for the odd sites to obtain X_{n+1}. Then $(X_n)_{n \geq 0}$ is a Markov chain in I^+ with invariant distribution $\pi(\cdot \mid y)$.

6
Appendix: probability and measure

Section 6.1 contains some reminders about countable sets and the discrete version of measure theory. For much of the book we can do without explicit mention of more general aspects of measure theory, except an elementary understanding of Riemann integration or Lebesgue measure. This is because the state-space is at worst countable. The proofs we have given may be read on two levels, with or without a measure-theoretic background. When interpreted in terms of measure theory, the proofs are intended to be rigorous. The basic framework of measure and probability is reviewed in Sections 6.2 and 6.3. Two important results of measure theory, the monotone convergence theorem and Fubini's theorem, are needed a number of times: these are discussed in Section 6.4. One crucial result which we found impossible to discuss convincingly without measure theory is the strong Markov property for continuous-time chains. This is proved in Section 6.5. Finally, in Section 6.6, we discuss a general technique for determining probability measures and independence in terms of π-systems, which are often more convenient than σ-algebras.

6.1 Countable sets and countable sums

A set I is *countable* if there is a bijection $f : \{1, \ldots, n\} \to I$ for some $n \in \mathbb{N}$, or a bijection $f : \mathbb{N} \to I$. In either case we can *enumerate* all the elements of I

$$f_1, f_2, f_3, \ldots$$

where in one case the sequence terminates and in the other it does not. There would have been no loss in generality had we insisted that all our Markov chains had state-space $\mathbb{N}$ or $\{1,\dots,n\}$ for some $n \in \mathbb{N}$: this just corresponds to a particular choice of the bijection f.

Any subset of a countable set is countable. Any finite cartesian product of countable sets is countable, for example $\mathbb{Z}^n$ for any n. Any countable union of countable sets is countable. The set of all subsets of $\mathbb{N}$ is uncountable and so is the set of real numbers $\mathbb{R}$.

We need the following basic fact.

Lemma 6.1.1. *Let I be a countably infinite set and let $\lambda_i \geq 0$ for all $i \in I$. Then, for any two enumerations of I*

$$i_1, i_2, i_3, \dots,$$
$$j_1, j_2, j_3, \dots$$

we have

$$\sum_{n=1}^{\infty} \lambda_{i_n} = \sum_{n=1}^{\infty} \lambda_{j_n}.$$

Proof. Given any $N \in \mathbb{N}$ we can find $M \geq N$ and $N' \geq M$ such that

$$\{i_1, \dots, i_N\} \subseteq \{j_1, \dots, j_M\} \subseteq \{i_1, \dots, i_{N'}\}.$$

Then

$$\sum_{n=1}^{N} \lambda_{i_n} \leq \sum_{n=1}^{M} \lambda_{j_n} \leq \sum_{n=1}^{N'} \lambda_{i_n}$$

and the result follows on letting $N \to \infty$. □

Since the value of the sum does not depend on the enumeration we are justified in using a notation which does not specify an enumeration and write simply

$$\sum_{i \in I} \lambda_i.$$

More generally, if we allow λ_i to take negative values, then we can set

$$\sum_{i \in I} \lambda_i = \left(\sum_{i \in I^+} \lambda_i\right) - \left(\sum_{i \in I^-} (-\lambda_i)\right)$$

where

$$I^\pm = \{i \in I : \pm\lambda_i \geq 0\},$$

allowing that the sum over I is undefined when the sums over I^+ and I^- are both infinite. There is no difficulty in showing for $\lambda_i, \mu_i \geq 0$ that

$$\sum_{i \in I}(\lambda_i + \mu_i) = \sum_{i \in I}\lambda_i + \sum_{i \in I}\mu_i.$$

By induction, for any finite set J and for $\lambda_{ij} \geq 0$, we have

$$\sum_{i \in I}\left(\sum_{j \in J}\lambda_{ij}\right) = \sum_{j \in J}\left(\sum_{i \in I}\lambda_{ij}\right).$$

The following two results on sums are simple versions of fundamental results for integrals. We take the opportunity to prove these simple versions in order to convey some intuition relevant to the general case.

Lemma 6.1.2 (Fubini's theorem – discrete case). *Let I and J be countable sets and let $\lambda_{ij} \geq 0$ for all $i \in I$ and $j \in J$. Then*

$$\sum_{i \in I}\left(\sum_{j \in J}\lambda_{ij}\right) = \sum_{j \in J}\left(\sum_{i \in I}\lambda_{ij}\right).$$

Proof. Let $j_1, j_2, j_3, \dots$ be an enumeration of J. Then

$$\sum_{i \in I}\left(\sum_{j \in J}\lambda_{ij}\right) \geq \sum_{i \in I}\left(\sum_{k=1}^{n}\lambda_{ij_k}\right) = \sum_{k=1}^{n}\left(\sum_{i \in I}\lambda_{ij_k}\right) \uparrow \sum_{j \in J}\left(\sum_{i \in I}\lambda_{ij}\right)$$

as $n \to \infty$. Hence

$$\sum_{i \in I}\left(\sum_{j \in J}\lambda_{ij}\right) \geq \sum_{j \in J}\left(\sum_{i \in I}\lambda_{ij}\right)$$

and the result follows by symmetry. □

Lemma 6.1.3 (Monotone convergence – discrete case). *Suppose for each $i \in I$ we are given an increasing sequence $(\lambda_i(n))_{n \geq 0}$ with limit λ_i, and that $\lambda_i(n) \geq 0$ for all i and n. Then*

$$\sum_{i \in I}\lambda_i(n) \uparrow \sum_{i \in I}\lambda_i \quad \text{as } n \to \infty.$$

Proof. Set $\delta_i(1) = \lambda_i(1)$ and for $n \geq 2$ set
$$\delta_i(n) = \lambda_i(n) - \lambda_i(n-1).$$
Then $\delta_i(n) \geq 0$ for all i and n, so as $n \to \infty$, by Fubini's theorem
$$\sum_{i \in I} \lambda_i(n) = \sum_{i \in I} \left(\sum_{k=1}^{n} \delta_i(k) \right)$$
$$= \sum_{k=1}^{n} \left(\sum_{i \in I} \delta_i(k) \right) \uparrow \sum_{k=1}^{\infty} \left(\sum_{i \in I} \delta_i(k) \right)$$
$$= \sum_{i \in I} \left(\sum_{k=1}^{\infty} \delta_i(k) \right) = \sum_{i \in I} \lambda_i. \qquad \square$$

6.2 Basic facts of measure theory

We state here for easy reference the basic definitions and results of measure theory. Let E be a set. A *σ-algebra* $\mathcal{E}$ on E is a set of subsets of E satisfying

(i) $\emptyset \in \mathcal{E}$;
(ii) $A \in \mathcal{E} \Rightarrow A^c \in \mathcal{E}$;
(iii) $(A_n \in \mathcal{E}, n \in \mathbb{N}) \Rightarrow \bigcup_n A_n \in \mathcal{E}$.

Here A^c denotes the complement $E \setminus A$ of A in E. Thus $\mathcal{E}$ is closed under countable set operations. The pair $(E, \mathcal{E})$ is called a *measurable space*. A *measure* μ on $(E, \mathcal{E})$ is a function $\mu : \mathcal{E} \to [0, \infty]$ which has the following *countable additivity* property:

$$(A_n \in \mathcal{E}, n \in \mathbb{N}, A_n \text{ disjoint}) \Rightarrow \mu\left(\bigcup_n A_n\right) = \sum_n \mu(A_n).$$

The triple $(E, \mathcal{E}, \mu)$ is called a *measure space*. If there exist sets $E_n \in \mathcal{E}$, $n \in \mathbb{N}$ with $\bigcup_n E_n = E$ and $\mu(E_n) < \infty$ for all n, then we say μ is *σ-finite*.

Example 6.2.1

Let I be a countable set and denote by $\mathcal{I}$ the set of all subsets of I. Recall that $\lambda = (\lambda_i : i \in I)$ is a measure in the sense of Section 1.1 if $\lambda_i \in [0, \infty)$ for all i. For such λ we obtain a measure on the measurable space $(I, \mathcal{I})$ by setting
$$\lambda(A) = \sum_{i \in A} \lambda_i.$$

In fact, we obtain in this way all σ-finite measures μ on $(I, \mathcal{I})$.

Example 6.2.2

Let $\mathcal{A}$ be any set of subsets of E. The set of all subsets of E is a σ-algebra containing $\mathcal{A}$. The intersection of any collection of σ-algebras is again a σ-algebra. The collection of σ-algebras containing $\mathcal{A}$ is therefore non-empty and its intersection is a σ-algebra $\sigma(\mathcal{A})$, which is called the σ-algebra *generated by* $\mathcal{A}$.

Example 6.2.3

In the preceding example take $E = \mathbb{R}$ and

$$\mathcal{A} = \{(a,b) : a,b \in \mathbb{R}, a < b\}.$$

The σ-algebra $\mathcal{B}$ generated by $\mathcal{A}$ is called the *Borel σ-algebra* of $\mathbb{R}$. It can be shown that there is a unique measure μ on $(\mathbb{R}, \mathcal{B})$ such that

$$\mu(a,b) = b - a \quad \text{for all } a, b.$$

This measure μ is called *Lebesgue measure*.

Let $(E_1, \mathcal{E}_1)$ and $(E_2, \mathcal{E}_2)$ be measurable spaces. A function $f : E_1 \to E_2$ is *measurable* if $f^{-1}(A) \in \mathcal{E}_1$ whenever $A \in \mathcal{E}_2$. When the range $E_2 = \mathbb{R}$ we take $\mathcal{E}_2 = \mathcal{B}$ by default. When the range E_2 is a countable set I we take $\mathcal{E}_2$ to be the set of all subsets $\mathcal{I}$ by default.

Let $(E, \mathcal{E})$ be a measurable space. We denote by $m\mathcal{E}$ the set of measurable functions $f : E \to \mathbb{R}$. Then $m\mathcal{E}$ is a vector space. We denote by $m\mathcal{E}^+$ the set of measurable functions $f : E \to [0, \infty]$, where we take on $[0, \infty]$ the σ-algebra generated by the open intervals (a, b). Then $m\mathcal{E}^+$ is a cone

$$(f, g \in m\mathcal{E}^+, \alpha, \beta \geq 0) \Rightarrow \alpha f + \beta g \in m\mathcal{E}^+.$$

Also, $m\mathcal{E}^+$ is closed under countable suprema:

$$(f_i \in m\mathcal{E}^+, i \in I) \Rightarrow \sup_i f_i \in m\mathcal{E}^+.$$

It follows that, for a sequence of functions $f_n \in m\mathcal{E}^+$, both $\limsup_n f_n$ and $\liminf_n f_n$ are in $m\mathcal{E}^+$, and so is $\lim_n f_n$ when this exists. It can be shown that there is a unique map $\widetilde{\mu} : m\mathcal{E}^+ \to [0, \infty]$ such that

(i) $\widetilde{\mu}(1_A) = \mu(A)$ for all $A \in \mathcal{E}$;
(ii) $\widetilde{\mu}(\alpha f + \beta g) = \alpha \widetilde{\mu}(f) + \beta \widetilde{\mu}(g)$ for all $f, g \in m\mathcal{E}^+, \alpha, \beta \geq 0$;
(iii) $(f_n \in m\mathcal{E}^+, n \in \mathbb{N}) \Rightarrow \widetilde{\mu}(\sum_n f_n) = \sum_n \widetilde{\mu}(f_n)$.

For $f \in m\mathcal{E}$, set $f^{\pm} = (\pm f) \vee 0$, then $f^+, f^- \in m\mathcal{E}^+$, $f = f^+ - f^-$ and $|f| = f^+ + f^-$. If $\tilde{\mu}(|f|) < \infty$ then f is said to be *integrable* and we set

$$\tilde{\mu}(f) = \mu(f^+) - \mu(f^-).$$

We call $\tilde{\mu}(f)$ the *integral* of f. It is conventional to drop the tilde and denote the integral by one of the following alternative notations:

$$\mu(f) = \int_E f d\mu = \int_{x \in E} f(x)\mu(dx).$$

In the case of Lebesgue measure μ, one usually writes simply

$$\int_{x \in \mathbb{R}} f(x) dx.$$

6.3 Probability spaces and expectation

The basic apparatus for modelling randomness is a *probability space* $(\Omega, \mathcal{F}, \mathbb{P})$. This is simply a measure space with total mass $\mathbb{P}(\Omega) = 1$. Thus $\mathcal{F}$ is a σ-algebra of subsets of Ω and $\mathbb{P} : \mathcal{F} \to [0, 1]$ satisfies

(i) $\mathbb{P}(\Omega) = 1$;
(ii) $\mathbb{P}(A_1 \cap A_2) = \mathbb{P}(A_1) + \mathbb{P}(A_2)$ for A_1, A_2 disjoint;
(iii) $\mathbb{P}(A_n) \uparrow \mathbb{P}(A)$ whenever $A_n \uparrow A$.

In (iii) we write $A_n \uparrow A$ to mean $A_1 \subseteq A_n \subseteq \ldots$ with $\bigcup_n A_n = A$. A measurable function X defined on $(\Omega, \mathcal{F})$ is called a *random variable*. We use random variables $Y : \Omega \to \mathbb{R}$ to model random quantities, where for a Borel set $B \subseteq \mathbb{R}$ the probability that $Y \in B$ is given by

$$\mathbb{P}(Y \in B) = \mathbb{P}(\{\omega : Y(\omega) \in B\}).$$

Similarly, given a countable state-space I, a random variable $X : \Omega \to I$ models a random state, with *distribution*

$$\lambda_i = \mathbb{P}(X = i) = \mathbb{P}(\{\omega : X(\omega) = i\}).$$

To every non-negative or integrable real-valued random variable Y is associated an average value or *expectation* $\mathbb{E}(Y)$, which is the integral of Y with respect to $\mathbb{P}$. Thus we have

(i) $\mathbb{E}(1_A) = \mathbb{P}(A)$ for $A \in \mathcal{F}$;
(ii) $\mathbb{E}(\alpha X + \beta Y) = \alpha \mathbb{E}(X) + \beta \mathbb{E}(Y)$ for $X, Y \in m\mathcal{F}^+$, $\alpha, \beta \geq 0$;

(iii) $(Y_n \in m\mathcal{F}^+, n \in \mathbb{N}, Y_n \uparrow Y) \Rightarrow \mathbb{E}(Y_n) \uparrow \mathbb{E}(Y)$.

When X is a random variable with values in I and $f : I \to [0, \infty]$ the expectation of $Y = f(X) = f \circ X$ is given explicitly by

$$\mathbb{E}(f(X)) = \sum_{i \in I} \lambda_i f_i$$

where λ is the distribution of X. For a real-valued random variable Y the probabilities are sometimes given by a measurable *density* function ρ in terms of Lebesgue measure:

$$\mathbb{P}(Y \in B) = \int_B \rho(y) dy.$$

Then for any measurable function $f : \mathbb{R} \to [0, \infty]$ there is an explicit formula

$$\mathbb{E}(f(Y)) = \int_{\mathbb{R}} f(y) \rho(y) dy.$$

6.4 Monotone convergence and Fubini's theorem

Here are the two theorems from measure theory that come into play in the main text. First we shall state the theorems, then we shall discuss some places where they are used. Proofs may be found, for example, in *Probability with Martingales* by D. Williams (Cambridge University Press, 1991).

Theorem 6.4.1 (Monotone convergence). Let $(E, \mathcal{E}, \mu)$ be a measure space and let $(f_n)_{n \geq 1}$ be a sequence of non-negative measurable functions. Then, as $n \to \infty$

$$(f_n(x) \uparrow f(x) \text{ for all } x \in E) \Rightarrow \mu(f_n) \uparrow \mu(f).$$

Theorem 6.4.2 (Fubini's theorem). Let $(E_1, \mathcal{E}_1, \mu_1)$ and $(E_2, \mathcal{E}_2, \mu_2)$ be two σ-finite measure spaces. Suppose that $f : E_1 \times E_2 \to [0, \infty]$ satisfies
   (i) $x \mapsto f(x, y) : E_1 \to [0, \infty]$ is $\mathcal{E}_1$ measurable for all $y \in E_2$;
   (ii) $y \mapsto \int_{x \in E_1} f(x, y) \mu_1(dx) : E_2 \to [0, \infty]$ is $\mathcal{E}_2$ measurable.
Then
   (a) $y \mapsto f(x, y) : E_2 \to [0, \infty]$ is $\mathcal{E}_2$ measurable for all $x \in E_1$;
   (b) $x \mapsto \int_{y \in E_2} f(x, y) \mu_2(dy) : E_1 \to [0, \infty]$ is $\mathcal{E}_1$ measurable;
   (c) $\int_{x \in E_1} \left(\int_{y \in E_2} f(x, y) \mu_2(dy) \right) \mu_1(dx) = \int_{y \in E_2} \left(\int_{x \in E_1} f(x, y) \mu_1(dx) \right) \mu_2(dy).$

The measurability conditions in the above theorems rarely need much consideration. They are powerful results and very easy to use. There is an equivalent formulation of monotone convergence in terms of sums: for non-negative measurable functions g_n we have

$$\mu\left(\sum_{n=1}^{\infty} g_n\right) = \sum_{n=1}^{\infty} \mu(g_n).$$

To see this just take $f_n = g_1 + \cdots + g_n$. This form of monotone convergence has already appeared in Section 6.2 as a defining property of the integral. This is also a special case of Fubini's theorem, provided that $(E, \mathcal{E}, \mu)$ is σ-finite: just take $E_2 = \{1, 2, 3, \dots\}$ and $\mu_2(\{n\}) = 1$ for all n.

We used monotone convergence in Theorem 1.10.1 to see that for a non-negative random variable Y we have

$$\mathbb{E}(Y) = \lim_{N \to \infty} \mathbb{E}(Y \wedge N).$$

We used monotone convergence in Theorem 2.3.2 to see that for random variables $S_n \geq 0$ we have

$$\mathbb{E}\left(\sum_n S_n\right) = \sum_n \mathbb{E}(S_n)$$

and

$$\mathbb{E}\left(\exp\left\{-\sum_n S_n\right\}\right) = \mathbb{E}\left(\lim_{N \to \infty} \exp\left\{-\sum_{n \leq N} S_n\right\}\right)$$
$$= \lim_{N \to \infty} \mathbb{E}\left(\exp\left\{-\sum_{n \leq N} S_n\right\}\right).$$

In the last application convergence is not monotone increasing but monotone decreasing. But if $0 \leq X_n \leq Y$ and $X_n \downarrow X$ then $Y - X_n \uparrow Y - X$. So $\mathbb{E}(Y - X_n) \uparrow \mathbb{E}(Y - X)$ and if $\mathbb{E}(Y) < \infty$ we can deduce $\mathbb{E}(X_n) \downarrow \mathbb{E}(X)$.

Fubini's theorem is used in Theorem 3.4.2 to see that

$$\int_0^{\infty} p_{ii}(t) dt = \int_0^{\infty} \mathbb{E}_i\left(1_{\{X_t = i\}}\right) dt = \mathbb{E}_i \int_0^{\infty} 1_{\{X_t = i\}} dt.$$

Thus we have taken $(E_1, \mathcal{E}_1, \mu_1)$ to be $[0, \infty)$ with Lebesgue measure and $(E_2, \mathcal{E}_2, \mu_2)$ to be the probability space with the measure $\mathbb{P}_i$.

6.5 Stopping times and the strong Markov property

The strong Markov property for continuous-time Markov chains cannot properly be understood without measure theory. The problem lies with

6.5 Stopping times and the strong Markov property

the notion of 'depending only on', which in measure theory is made precise as measurability with respect to some σ-algebra. Without measure theory the statement that a set A depends only on $(X_s : s \leq t)$ does not have a precise meaning. Of course, if the dependence is reasonably explicit we can exhibit it, but then, in general, in what terms would you require the dependence to be exhibited? So in this section we shall give a precise measure-theoretic account of the strong Markov property.

Let $(X_t)_{t \geq 0}$ be a right-continuous process with values in a countable set I. Denote by $\mathcal{F}_t$ the σ-algebra generated by $\{X_s : s \leq t\}$, that is to say, by all sets $\{X_s = i\}$ for $s \leq t$ and $i \in I$. We say that a random variable T with values in $[0, \infty]$ is a *stopping time* of $(X_t)_{t \geq 0}$ if $\{T \leq t\} \in \mathcal{F}_t$ for all $t \geq 0$. Note that this certainly implies

$$\{T < t\} = \bigcup_n \{T \leq t - 1/n\} \in \mathcal{F}_t \quad \text{for all} \quad t \geq 0.$$

We define for stopping times T

$$\mathcal{F}_T = \{A \in \mathcal{F} : A \cap \{T \leq t\} \in \mathcal{F}_t \text{ for all } t \geq 0\}.$$

This turns out to be the correct way to make precise the notion of sets which 'depend only on $\{X_t : t \leq T\}$'.

Lemma 6.5.1. *Let S and T be stopping times of $(X_t)_{t \geq 0}$. Then both X_T and $\{S \leq T\}$ are $\mathcal{F}_T$-measurable.*

Proof. Since $(X_t)_{t \geq 0}$ is right-continuous, on $\{T < t\}$ there exists an $n \geq 0$ such that for all $m \geq n$, for some $k \geq 1$, $(k-1)2^{-m} \leq T < k2^{-m} \leq t$ and $X_{k2^{-m}} = X_T$. Hence

$$\{X_T = i\} \cap \{T \leq t\}$$
$$= (\{X_t = i\} \cap \{T = t\})$$
$$\cup \left(\bigcup_{n=0}^{\infty} \bigcap_{m=n}^{\infty} \bigcup_{k=1}^{[2^m t]} \{X_{k2^{-m}} = i\} \cap \{(k-1)2^{-m} \leq T < k2^{-m}\} \right) \in \mathcal{F}_t$$

so X_T is $\mathcal{F}_T$-measurable.

We have

$$\{S > T\} \cap \{T \leq t\} = \bigcup_{s \in \mathbb{Q}, s \leq t} (\{T \leq s\} \cap \{S > s\}) \in \mathcal{F}_t$$

so $\{S > T\} \in \mathcal{F}_T$, and so $\{S \leq T\} \in \mathcal{F}_T$. □

Lemma 6.5.2. For all $m \geq 0$, the jump time J_m is a stopping time of $(X_t)_{t \geq 0}$.

Proof. Obviously, $J_0 = 0$ is a stopping time. Assume inductively that J_m is a stopping time. Then
$$\{J_{m+1} \leq t\} = \bigcup_{s \in \mathbb{Q}, s \leq t} \{J_m \leq s\} \cap \{X_s \neq X_{J_m}\} \in \mathcal{F}_t$$
for all $t \geq 0$, so J_{m+1} is a stopping time and the induction proceeds. □

We denote by $\mathcal{G}_m$ the σ-algebra generated by $Y_0, \ldots, Y_m$ and $S_1, \ldots, S_m$, that is, by events of the form $\{Y_k = i\}$ for $k \leq m$ and $i \in I$ or of the form $\{S_k > s\}$ for $k \leq m$ and $s > 0$.

Lemma 6.5.3. Let T be a stopping time of $(X_t)_{t \geq 0}$ and let $A \in \mathcal{F}_T$. Then for all $m \geq 0$ there exist a random variable T_m and a set A_m, both measurable with respect to $\mathcal{G}_m$, such that $T = T_m$ and $1_A = 1_{A_m}$ on $\{T < J_{m+1}\}$.

Proof. Fix $t \geq 0$ and consider
$$\mathcal{A}_t = \{A \in \mathcal{F}_t : A \cap \{t < J_{m+1}\} = A_m \cap \{t < J_{m+1}\} \quad \text{for some } A_m \in \mathcal{G}_m\}.$$
Since $\mathcal{G}_m$ is a σ-algebra, so is $\mathcal{A}_t$. For $s \leq t$ we have
$$\{X_s = i\} \cap \{t < J_{m+1}\}$$
$$= \left(\bigcup_{k=0}^{m-1} \{Y_k = i, J_k \leq s < J_{k+1}\} \cup \{Y_m = i, J_m \leq s\} \right) \cap \{t < J_{m+1}\}$$
so $\{X_s = i\} \in \mathcal{A}_t$. Since these sets generate $\mathcal{F}_t$, this implies that $\mathcal{A}_t = \mathcal{F}_t$.

For T a stopping time and $A \in \mathcal{F}_T$ we have $B(t) := \{T \leq t\} \in \mathcal{F}_t$ and $A(t) := A \cap \{T \leq t\} \in \mathcal{F}_t$ for all $t \geq 0$. So we can find $B_m(t), A_m(t) \in \mathcal{G}_m$ such that
$$B(t) \cap \{T < J_{m+1}\} = B_m(t) \cap \{T < J_{m+1}\},$$
$$A(t) \cap \{T < J_{m+1}\} = A_m(t) \cap \{T < J_{m+1}\}.$$
Set
$$T_m = \sup_{t \in \mathbb{Q}} t 1_{B_m(t)}, \quad A_m = \bigcup_{t \in \mathbb{Q}} A_m(t)$$
then T_m and A_m are $\mathcal{G}_m$-measurable and
$$T_m 1_{\{T < J_{m+1}\}} = \sup_{t \in \mathbb{Q}} t 1_{B_m(t) \cap \{T < J_{m+1}\}}$$
$$= \left(\sup_{t \in \mathbb{Q}} t 1_{\{T \leq t\}} \right) 1_{\{T < J_{m+1}\}} = T 1_{\{T < J_{m+1}\}}$$

6.5 Stopping times and the strong Markov property 227

and

$$A_m \cap \{T < J_{m+1}\} = \bigcup_{t \in \mathbb{Q}} A_m(t) \cap \{T < J_{m+1}\}$$
$$= \bigcup_{t \in \mathbb{Q}} (A \cap \{T \le t\}) \cap \{T < J_{m+1}\} = A \cap \{T < J_{m+1}\}$$

as required. □

Theorem 6.5.4 (Strong Markov property). Let $(X_t)_{t \ge 0}$ be Markov(λ, Q) and let T be a stopping time of $(X_t)_{t \ge 0}$. Then, conditional on $T < \zeta$ and $X_T = i$, $(X_{T+t})_{t \ge 0}$ is Markov(δ_i, Q) and independent of $\mathcal{F}_T$.

Proof. On $\{T < \zeta\}$ set $\tilde{X}_t = X_{T+t}$ and denote by $(\tilde{Y}_n)_{n \ge 0}$ the jump chain and by $(\tilde{S}_n)_{n \ge 1}$ the holding times of $(\tilde{X}_t)_{t \ge 0}$. We have to show that, for all $A \in \mathcal{F}_T$, all $i_0, \ldots, i_n \in I$ and all $s_1, \ldots, s_n \ge 0$

$$\mathbb{P}\big(\{\tilde{Y}_0 = i_0, \ldots, \tilde{Y}_n = i_n, \tilde{S}_1 > s_1, \ldots, \tilde{S}_n > s_n\} \cap A \cap \{T < \zeta\} \cap \{X_T = i\}\big)$$
$$= \mathbb{P}_i(Y_0 = i_0, \ldots, Y_n = i_n, S_1 > s_1, \ldots, S_n > s_n)$$
$$\times \mathbb{P}\big(A \cap \{T < \zeta\} \cap \{X_T = i\}\big).$$

It suffices to prove this with $\{T < \zeta\}$ replaced by $\{J_m \le T < J_{m+1}\}$ for all $m \ge 0$ and then sum over m. By Lemmas 6.5.1 and 6.5.2, $\{J_m \le T\} \cap \{X_T = i\} \in \mathcal{F}_T$ so we may assume without loss of generality that $A \subseteq \{J_m \le T\} \cap \{X_T = i\}$. By Lemma 6.5.3 we can write $T = T_m$ and $1_A = 1_{A_m}$ on $\{T < J_{m+1}\}$, where T_m and A_m are $\mathcal{G}_m$-measurable.

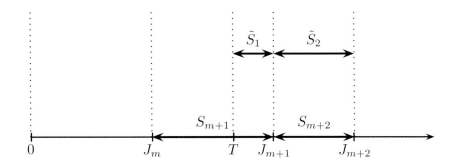

On $\{J_m \le T < J_{m+1}\}$ we have, as shown in the diagram

$$\tilde{Y}_n = Y_{m+n}, \quad \tilde{S}_1 = S_{m+1} - (T - J_m), \quad \tilde{S}_n = S_{m+n} \text{ for } n \ge 2.$$

Now, conditional on $Y_m = i$, S_{m+1} is independent of $\mathcal{G}_m$ and hence of $T_m - J_m$ and A_m and, by the memoryless property of the exponential

$$\mathbb{P}\big(S_{m+1} > s_1 + (T_m - J_m) \mid Y_m = i, S_{m+1} > T_m - J_m\big) = e^{-q_i s_1} = \mathbb{P}_i(S_1 > s_1).$$

Hence, by the Markov property of the jump chain

$$\begin{aligned}
\mathbb{P}\big(\{\widetilde{Y}_0 = i_0, \ldots, &\widetilde{Y}_n = i_n, \\
\widetilde{S}_1 > s_1, \ldots, &\widetilde{S}_n > s_n\} \cap A \cap \{J_m \leq T < J_{m+1}\} \cap \{X_T = i\}\big) \\
= \mathbb{P}\big(\{Y_m = i_0, &\ldots, Y_{m+n} = i_n, S_{m+1} > s_1 + (T_m - J_m), \\
S_{m+2} > s_2, &\ldots, S_{m+n} > s_n\} \cap A_m \cap \{S_{m+1} > T_m - J_m\}\big) \\
= \mathbb{P}_i(Y_0 = i_0, &\ldots, Y_n = i_n, \\
S_1 > s_1, &\ldots, S_n > s_n) \mathbb{P}\big(A \cap \{J_m \leq T < J_{m+1}\} \cap \{X_T = i\}\big)
\end{aligned}$$

as required. □

6.6 Uniqueness of probabilities and independence of σ-algebras

For both discrete-time and continuous-time Markov chains we have given definitions which specify the probabilities of certain events determined by the process. From these specified probabilities we have often deduced explicitly the values of other probabilities, for example hitting probabilities. In this section we shall show, in measure-theoretic terms, that our definitions determine the probabilities of *all* events depending on the process. The constructive approach we have taken should make this seem obvious, but it is illuminating to see what has to be done.

Let Ω be a set. A π-*system* $\mathcal{A}$ on Ω is a collection of subsets of Ω which is closed under finite intersections; thus

$$A_1, A_2 \in \mathcal{A} \Rightarrow A_1 \cap A_2 \in \mathcal{A}.$$

We denote as usual by $\sigma(\mathcal{A})$ the σ-algebra generated by $\mathcal{A}$. If $\sigma(\mathcal{A}) = \mathcal{F}$ we say that $\mathcal{A}$ *generates* $\mathcal{F}$.

Theorem 6.6.1. *Let $(\Omega, \mathcal{F})$ be a measurable space. Let $\mathbb{P}_1$ and $\mathbb{P}_2$ be probability measures on $(\Omega, \mathcal{F})$ which agree on a π-system $\mathcal{A}$ generating $\mathcal{F}$. Then $\mathbb{P}_1 = \mathbb{P}_2$.*

Proof. Consider
$$\mathcal{D} = \{A \in \mathcal{F} : \mathbb{P}_1(A) = \mathbb{P}_2(A)\}.$$

6.6 Uniqueness of probabilities and independence of σ-algebras

We have assumed that $\mathcal{A} \subseteq \mathcal{D}$. Moreover, since $\mathbb{P}_1$ and $\mathbb{P}_2$ are probability measures, $\mathcal{D}$ has the following properties:

(i) $\Omega \in \mathcal{D}$;
(ii) $(A, B \in \mathcal{D}$ and $A \subseteq B) \Rightarrow B \backslash A \in \mathcal{D}$;
(iii) $(A_n \in \mathcal{D}, A_n \uparrow A) \Rightarrow A \in \mathcal{D}$.

Any collection of subsets having these properties is called a *d-system*. Since $\mathcal{A}$ generates $\mathcal{F}$, the result now follows from the following lemma. $\square$

Lemma 6.6.2 (Dynkin's π-system lemma). *Let $\mathcal{A}$ be a π-system and let $\mathcal{D}$ be a d-system. Suppose $\mathcal{A} \subseteq \mathcal{D}$. Then $\sigma(\mathcal{A}) \subseteq \mathcal{D}$.*

Proof. Any intersection of d-systems is again a d-system, so we may without loss assume that $\mathcal{D}$ is the smallest d-system containing $\mathcal{A}$. You may easily check that any d-system which is also a π-system is necessarily a σ-algebra, so it suffices to show $\mathcal{D}$ is a π-system. This we do in two stages.

Consider first
$$\mathcal{D}_1 = \{A \in \mathcal{D} : A \cap B \in \mathcal{D} \text{ for all } B \in \mathcal{A}\}.$$

Since $\mathcal{A}$ is a π-system, $\mathcal{A} \subseteq \mathcal{D}_1$. You may easily check that $\mathcal{D}_1$ is a d-system – because $\mathcal{D}$ is a d-system. Since $\mathcal{D}$ is the smallest d-system containing $\mathcal{A}$, this shows $\mathcal{D}_1 = \mathcal{D}$.

Next consider
$$\mathcal{D}_2 = \{A \in \mathcal{D} : A \cap B \in \mathcal{D} \text{ for all } B \in \mathcal{D}\}.$$

Since $\mathcal{D}_1 = \mathcal{D}$, $\mathcal{A} \subseteq \mathcal{D}_2$. You can easily check that $\mathcal{D}_2$ is also a d-system. Hence also $\mathcal{D}_2 = \mathcal{D}$. But this shows $\mathcal{D}$ is a π-system. $\square$

The notion of independence used in advanced probability is the independence of σ-algebras. Suppose that $(\Omega, \mathcal{F}, \mathbb{P})$ is a probability space and $\mathcal{F}_1$ and $\mathcal{F}_2$ are sub-σ-algebras of $\mathcal{F}$. We say that $\mathcal{F}_1$ and $\mathcal{F}_2$ are *independent* if
$$\mathbb{P}(A_1 \cap A_2) = \mathbb{P}(A_1)\mathbb{P}(A_2) \quad \text{for all } A_1 \in \mathcal{F}_1, A_2 \in \mathcal{F}_2.$$

The usual means of establishing such independence is the following corollary of Theorem 6.6.1.

Theorem 6.6.3. *Let $\mathcal{A}_1$ be a π-system generating $\mathcal{F}_1$ and let $\mathcal{A}_2$ be a π-system generating $\mathcal{F}_2$. Suppose that*
$$\mathbb{P}(A_1 \cap A_2) = \mathbb{P}(A_1)\mathbb{P}(A_2) \quad \text{for all } A_1 \in \mathcal{A}_1, A_2 \in \mathcal{A}_2.$$

Then $\mathcal{F}_1$ and $\mathcal{F}_2$ are independent.

Proof. There are two steps. First fix $A_2 \in \mathcal{A}_2$ with $\mathbb{P}(A_2) > 0$ and consider the probability measure

$$\widetilde{\mathbb{P}}(A) = \mathbb{P}(A \mid A_2).$$

We have assumed that $\widetilde{\mathbb{P}}(A) = \mathbb{P}(A)$ for all $A \in \mathcal{A}_1$, so, by Theorem 6.6.1, $\widetilde{\mathbb{P}} = \mathbb{P}$ on $\mathcal{F}_1$. Next fix $A_1 \in \mathcal{F}_1$ with $\mathbb{P}(A_1) > 0$ and consider the probability measure

$$\widetilde{\widetilde{\mathbb{P}}}(A) = \mathbb{P}(A \mid A_1).$$

We showed in the first step that $\widetilde{\widetilde{\mathbb{P}}}(A) = \mathbb{P}(A)$ for all $A \in \mathcal{A}_2$, so, by Theorem 6.6.1, $\widetilde{\widetilde{\mathbb{P}}} = \mathbb{P}$ on $\mathcal{F}_2$. Hence $\mathcal{F}_1$ and $\mathcal{F}_2$ are independent. □

We now review some points in the main text where Theorems 6.6.1 and 6.6.3 are relevant.

In Theorem 1.1.1 we showed that our definition of a discrete-time Markov chain $(X_n)_{n \geq 0}$ with initial distribution λ and transition matrix P determines the probabilities of all events of the form

$$\{X_0 = i_0, \ldots, X_n = i_n\}.$$

But subsequently we made explicit calculations for probabilities of events which were not of this form – such as the event that $(X_n)_{n \geq 0}$ visits a set of states A. We note now that the events $\{X_0 = i_0, \ldots, X_n = i_n\}$ form a π-system which generates the σ-algebra $\sigma(X_n : n \geq 0)$. Hence, by Theorem 6.6.1, our definition determines (in principle) the probabilities of all events in this σ-algebra.

In our general discussion of continuous-time random processes in Section 2.2 we claimed that for a right-continuous process $(X_t)_{t \geq 0}$ the probabilities of events of the form

$$\{X_{t_0} = i_0, \ldots, X_{t_n} = i_n\}$$

for all $n \geq 0$ determined the probabilities of all events depending on $(X_t)_{t \geq 0}$. Now events of the form $\{X_{t_0} = i_0, \ldots, X_{t_n} = i_n\}$ form a π-system which generates the σ-algebra $\sigma(X_t : t \geq 0)$. So Theorem 6.6.1 justifies (a precise version) of this claim. The point about right-continuity is that without such an assumption an event such as

$$\{X_t = i \text{ for some } t > 0\}$$

which might reasonably be considered to depend on $(X_t)_{t \geq 0}$, is *not necessarily measurable* with respect to $\sigma(X_t : t \geq 0)$. An argument given in

6.6 Uniqueness of probabilities and independence of σ-algebras 231

Section 2.2 shows that this event *is* measurable in the right-continuous case. We conclude that, without some assumption like right-continuity, general continuous-time processes are unreasonable.

Consider now the method of describing a minimal right-continuous process $(X_t)_{t\geq 0}$ via its jump process $(Y_n)_{n\geq 0}$ and holding times $(S_n)_{n\geq 1}$. Let us take $\mathcal{F} = \sigma(X_t : t \geq 0)$. Then Lemmas 6.5.1 and 6.5.2 show that $(Y_n)_{n\geq 0}$ and $(S_n)_{n\geq 1}$ are $\mathcal{F}$-measurable. Thus $\mathcal{G} \subseteq \mathcal{F}$ where

$$\mathcal{G} = \sigma\big((Y_n)_{n\geq 0}, (S_n)_{n\geq 1}\big).$$

On the other hand, for all $i \in I$

$$\{X_t = i\} = \bigcup_{n\geq 0} \{J_n \leq t < J_{n+1}\} \cap \{Y_n = i\} \in \mathcal{G},$$

so also $\mathcal{F} \subset \mathcal{G}$.

A useful π-system generating $\mathcal{G}$ is given by sets of the form

$$B = \{Y_0 = i_0, \ldots, Y_n = i_n, S_1 > s_1, \ldots, S_n > s_n\}.$$

Our jump chain/holding time definition of the continuous-time chain $(X_t)_{t\geq 0}$ with initial distribution λ and generator matrix Q may be read as stating that, for such events

$$\mathbb{P}(B) = \lambda_{i_0} \pi_{i_0 i_1} \ldots \pi_{i_{n-1} i_n} e^{-q_{i_0} s_1} \ldots e^{-q_{i_{n-1}} s_n}.$$

Then, by Theorem 6.6.1, this definition determines $\mathbb{P}$ on $\mathcal{G}$ and hence on $\mathcal{F}$.

Finally, we consider the strong Markov property, Theorem 6.5.4. Assume that $(X_t)_{t\geq 0}$ is Markov(λ, Q) and that T is a stopping time of $(X_t)_{t\geq 0}$. On the set $\tilde{\Omega} = \{T < \zeta\}$ define $\tilde{X}_t = X_{T+t}$ and let $\tilde{\mathcal{F}} = \sigma(\tilde{X}_t : t \geq 0)$; write $(\tilde{Y}_n)_{n\geq 0}$ and $(\tilde{S}_n)_{n\geq 0}$ for the jump chain and holding times of $(\tilde{X}_t)_{t\geq 0}$ and set

$$\tilde{\mathcal{G}} = \sigma\big((\tilde{Y}_n)_{n\geq 0}, (\tilde{S}_n)_{n\geq 0}\big).$$

Thus $\tilde{\mathcal{F}}$ and $\tilde{\mathcal{G}}$ are σ-algebras on $\tilde{\Omega}$, and coincide by the same argument as for $\mathcal{F} = \mathcal{G}$. Set

$$\tilde{B} = \{\tilde{Y}_0 = i_0, \ldots, \tilde{Y}_n = i_n, \tilde{S}_1 > s_1, \ldots, \tilde{S}_n > s_n\}.$$

Then the conclusion of the strong Markov property states that

$$\mathbb{P}(\tilde{B} \mid T < \zeta, X_T = i) = \mathbb{P}_i(B)$$

with B as above, and that

$$\mathbb{P}(\tilde{C} \cap A \mid T < \zeta, X_T = i) = \mathbb{P}(\tilde{C} \mid T < \zeta, X_t = i)\mathbb{P}(A \mid T < \zeta, X_T = i)$$

for all $\tilde{C} \in \tilde{\mathcal{F}}$ and $A \in \mathcal{F}_T$. By Theorem 6.6.3 it suffices to prove the independence assertion for the case $\tilde{C} = \tilde{B}$, which is what we did in the proof of Theorem 6.5.4.

Further reading

We gather here the references for further reading which have appeared in the text. This may provide a number of starting points for your exploration of the vast literature on Markov processes and their applications.

J. Besag, P. Green, D. Higdon and K. Mengersen, Bayesian computation and stochastic systems, *Statistical Science* **10 (1)** (1995), 3–40.

K.L. Chung, *Markov Chains with Stationary Transition Probabilities*, Springer, Berlin, 2nd edition, 1967.

P.G. Doyle and J.L. Snell, *Random Walks and Electrical Networks*, Carus Mathematical Monographs 22, Mathematical Association of America, 1984.

W.J. Ewens, *Mathematical Population Genetics*, Springer, Berlin, 1979.

D. Freedman, *Markov Chains*, Holden–Day, San Francisco, 1971.

W.R. Gilks, S. Richardson and D.J. Spiegelhalter, *Markov Chain Monte Carlo in Practice*, Chapman and Hall, London, 1996.

T.E. Harris, *The Theory of Branching Processes*, Dover, New York, 1989.

F.P. Kelly, *Reversibility and Stochastic Networks*, Wiley, Chichester, 1978.

D. Revuz, *Markov Chains*, North-Holland, Amsterdam, 1984.

B.D. Ripley, *Stochastic Simulation*, Wiley, Chichester, 1987.

L.C.G. Rogers and D. Williams, *Diffusions, Markov Processes and Martingales, Vol 1: Foundations*, Wiley, Chichester, 2nd edition, 1994.

S.M. Ross, *Applied Probability Models with Optimization Applications*, Holden–Day, San Francisco, 1970.

D.W. Stroock, *Probability Theory – An Analytic View*, Cambridge University Press, 1993.

H.C. Tijms, *Stochastic Models – an Algorithmic Approach*, Wiley, Chichester, 1994.

D. Williams, *Probability with Martingales*, Cambridge University Press, 1991.

Index

absorbing state 11, 111
absorption probability 12, 112
action 198
adapted 129
alleles 175
aperiodicity 40
average number of customers 181

backward equation 62, 96
Bayesian statistics 211
biological models 6, 9, 16, 82, 170
birth process 81
   infinitesimal definition 85
   jump chain/holding time definition 85
   transition probability definition 85
birth-and-death chain 16
boundary 138
boundary theory for Markov chains 147
branching process 171
   with immigration 179
Brownian motion 159
   as limit of random walks 164
   existence 161
   in $\mathbb{R}^d$ 165
   scaling invariance 165
   starting from x 165
   transition density 166
busy period 181

capacity 151
central limit theorem 160
charge 151
closed class 11, 111
closed migration process 184
communicating class 11, 111
conditional expectation 129
conductivity 151
continuous-time Markov chains 87
   construction 89
   infinitesimal definition 94
   jump chain/holding time definition 94, 97
   transition probability definition 94, 97
continuous-time random process 67
convergence to equilibrium 41, 121, 168
countable set 217
coupling method 41
current 151

detailed balance 48, 124
discounted value function 202
distribution 1

$\mathbb{E}$, expectation 222
$E(\lambda)$, exponential distribution of parameter λ 70
e^Q, exponential of Q 62
effective conductivity 156
electrical network 151
energy 154
epidemic model 173
equilibrium distribution 33, 117
ergodic theorem 53, 126, 168
Erlang's formula 183
excursions 24
expectation 222
expected hitting time 12, 113
expected return time 37, 118
explosion
    for birth processes 83
    for continuous-time chains 90
explosion time 69
explosive Q-matrix 91
exponential distribution 70

$\mathcal{F}_n$, filtration 129
fair price 135
filtration 129
finite-dimensional distributions 67
first passage decomposition 28
first passage time 19, 115
flow 151
forward equation 62, 100
    for birth processes 84
    for Poisson process 78
Fubini's theorem 223
    discrete case 219
full conditional distributions 212
fundamental solution 145

γ_i^j, expected time in i between visits to j 35
Galton–Watson process 171
gambling 131
Gaussian distribution 160
generator 166
generator matrix 94, 97

Gibbs sampler 210
gravity 134, 169
Green matrix 144, 145

harmonic function 146
Hastings algorithm 210
hitting probability 12
hitting time 12, 111
holding times 69

I, state-space 2
infective 173
integrable 129, 222
integral form of the backward equation 98
integral form of the forward equation 101
inter-arrival times 180
invariant distribution 33, 117
    computation of 40
irreducibility 11, 111
Ising model 214

jump chain 69
jump matrix 87
jump times 69

last exit time 20
long-run proportion of time 53, 126

m_i, expected return time 37, 118
μ_i^j, expected time in i between visits to j 118
Markov(λ, P) 2
Markov(λ, Q) 94, 97
Markov chain
    continuous-time 88
    discrete-time 2
Markov chain Monte Carlo 206, 208
Markov decision process 197
    expected total cost 198
    expected total discounted cost 202
    long-run average costs 204
Markov property 3
    for birth processes 84
    for continuous-time chains 93
    for Poisson process 75
martingale 129, 141, 176, 204

associated to a Markov chain 132
associated to Brownian motion 169
matrix exponentials 105
maximum likelihood estimate 56
measure 1
memoryless property 70
Metropolis algorithm 210
minimal non-negative solution 13
minimal process 69
monotone convergence 223
  discrete case 219
Moran model 177
mutation 6, 176

non-minimal chains 103
null recurrence 37, 118

$o(t), O(t)$, order notation 63
Ohm's law 151
open migration process 185
optional stopping theorem 130

$\mathbb{P}$, probability 222
P, transition matrix 2
$\widehat{P}$, transition matrix of reversed chain 47
$P(t)$, transition semigroup 96
$\widehat{P}(t)$, semigroup of reversed chain 124
$p_{ij}^{(n)}$, n-step transition probability 5
Π, jump matrix 87
π-system 228
Poisson process 74
  infinitesimal definition 76
  jump chain/holding time definition 76
  transition probability definition 76
policy 198
policy improvement 201
population genetics 175
population growth 171
positive recurrence 37, 118
potential 138
  associated to a Markov chain 138
  associated to Brownian motion 169

gravitational 134
in electrical networks 151
with discounted costs 142
potential theory 134
probability generating function 171
probability measure 222
probability space 222

Q-matrix 60
$\widehat{Q}$, generator matrix of reversed chain 124
q_i, rate of leaving i 61
q_{ij}, rate of going from i to j 61
queue 179
  M/G/1 187
  M/G/∞ 191
  M/M/1 180
  M/M/s 182
queueing network 183–185
queues in series 183

random chessboard knight 50
random walk
  on $\mathbb{Z}^d$ 29
  on a graph 49
recurrence 24, 114, 167
recurrence relations 57
reflected random walks 195
reservoir model 194, 195
resolvent 146
resource management 192
restocking a warehouse 192
return probability 25
reversibility 48, 125
right-continuous process 67
ruin
  gambler 15
  insurance company 196

selective advantage 176
semigroup 96
semigroup property 62
service times 180
shopping centre 185
simple birth process 82
simulation 206
skeleton 122
state-space 1

stationary distribution 33, 117
stationary increments 76
stationary policy 198
statistics 55, 211, 215
stochastic matrix 2
stopping time 19
strong law of large numbers 52
strong Markov property 19, 93, 227
success-run chain 38
susceptible 173

telephone exchange 183
texture parameter 216
time reversal 47, 123
transience 24, 114, 167
transition matrix 2
   irreducible 11
   maximum likelihood estimate 56
transition semigroup 165
truncated Poisson distribution 183

unit mass 3

$V_i(n)$, number of visits to i before n 53
valency 50
value function 198

weak convergence 164
Wiener process 159
Wiener's theorem 161
Wright–Fisher model 175

ζ, explosion time 69